Journey Through England

by Rob Neillands

GALLERY BOOKS
An Imprint of W. H. Smith Publishers Inc.
112 Madison Avenue
New York City 10016

Endpapers: Morris dancers at Chipping Campden in Gloucestershire
Title page: A tree-lined approach to Winchester Cathedral, Hampshire
Title verso/Introduction: Thaxted Guildhall and church, Essex
Contents page: The River Cam at Cambridge

Editor: Donna Wood
Art Editor: Edward Pitcher
Designer: Marion Neville
Maps: Eugene Fleury
Production: Steve Roberts

Produced by Marshall Cavendish Books Limited.
Published by
GALLERY BOOKS
An imprint of W.H. Smith Publishers Inc.
112 Madison Avenue, New York,
New York 10016

ISBN 0-8317-5260-2

Typeset in Palatino by TypeFast Ltd, London
Printed in Italy by L.E.G.O. S.p.a. Vicenza

Introduction

England's pretty, mild countryside never fails to charm the visitor. Seen from the air for the first time it is a picturebook land of fields and meadows hemmed together by hedgerows and broken up here and there by scattered villages, where it is taken for granted the cottages will be compounded of thatched roofs and black and white façades. This is, of course, quite a fair picture, but what about the wilder parts, the heady heights and plunging depths of the mountainous Lake District, the wave-lashed coast of Cornwall, the rugged cliffs, sandy coves, standing stones and ancient monuments that abound in this country, each with a past that stretches back into legend? England is far, far more than merely a picturesque place with a long history. It is full of fascinating nooks and crannies that need to be sought out, and then explored at leisure. This book leads you to these places, tells you the story behind them, points you in the right direction, then leaves you to make up your own mind. Organised into eight regions just the right size for touring, this book will be an invaluable companion throughout your Journey through England.

Contents

England – regional breakdown
1 The West Country
2 Southern England
3 London, Thames & Chilterns
4 East Anglia
5 The English Shires
8 The Northern Counties
Manchester
Birmingham
Norwich
Oxford
London
Southampton

BRISTOL CHANNEL
Lundy Island
Ilfracombe
Dunster
Exmoor
AVON
M 4
M 5
Marlborough
Avebury
Bath
WILTSHIRE
Wells
Stonehenge
Glastonbury
Stourhead
SOMERSET
Salisbury
A 30
A 303
A 39
Sherborne
DORSET
DEVON
Cerne Abbas
Athelhampton House
Exeter
Lyme Regis
Castle Drogo
A 35
A 38
Dartmoor
Corfe Castle
Bodmin Moor
Cotehele House
Newquay
Brixham
Plymouth
CORNWALL
Dartmouth
St Ives
Mousehole
St Michael's Mount
Helston
Land's End
0 10 20 30 40 km
0 10 20 miles

The West Country

The sweep of beach and the white chalk cliffs of Durdle Door, on the west coast of Dorset

RUNNING OUT INTO the Atlantic at Land's End, the West Country encapsulates much of the beauty and variety that is England. Even though they combine into a delightful whole, the counties of the West Country are distinctly different. Indeed, the true natives of Cornwall regard those of the British nation who live across the Tamar as 'foreigners', and say so at length. Cornwall is a county of small villages, a fey country, somewhat Celtic, with historic industries, like the mining of tin that dates back to Roman times. Devon is a maritime county, with the smack of the sea about it, the home of Drake and Raleigh, a wide, beautiful county, occupied in the centre by the vast wastes of Dartmoor.

Somerset is the rich cider county of England, a land awash with cream and honey, less known than Devon and Cornwall but not a place to miss. Among the local gems one can list the county town of Taunton and the pilgrim centre of Glastonbury.

The Dorset coastline is a beautiful, indented place, full of coves and bays and historic ports, with Lyme Regis as the Queen of them all. Inland, Corfe Castle, now in ruins, dominates the village below, an echo of the glory that is gone, while just to the north in the new county of Avon, Bath, that glorious Georgian city, continues to draw the crowds.

Finally, heading east, the traveller comes to windy Wiltshire, filled in the centre by Salisbury Plain, littered with historic churches and cathedrals, and the hilltop relics of the Iron Age.

ATHLEHAMPTON HOUSE
Dorset

Athlehampton House is yet another of those magnificent survivals from England's medieval past. Built for the Marlyn family in the 15th century, and much enlarged in the 16th, the Hall has a fine timbered roof, oriel windows with plenty of heraldic glass, and such intriguing features as hidden cupboards, secret stairways, and a Tudor chamber. The outside is equally attractive, for the house is surrounded by courtyards and walled gardens and is almost encircled by the river which acts as a moat. The stable block is thatched and the ten acres of garden contain a late-medieval dovecote. The last Marlyn died at the end of the 16th century, but subsequent owners have kept this house and the surrounding lands in the most beautiful condition, and not a lot has changed since this lovely house was built five centuries ago. The house is open to visitors on certain days.

ATHLEHAMPTON lies off the A35, 5 miles north-east of Dorchester.

A view through the gate across the garden to Athlehampton House

AVEBURY
Wiltshire

Avebury lies at the western end of the Ridgeway Path, a fittingly ancient terminal for this ancient trackway.

The village of Avebury is quite small, barely a hamlet, but although it is not as well known as Stonehenge, it contains the largest and most striking array of solitary *menhirs* (standing stones) and stone circles in Britain, dating from the Bronze Age. These encircle the village in serried ranks, most still standing, some sunken in the grass, the total effect very impressive. A short walk south from the centre brings the traveller to Silbury Hill, a pudding-basin shaped prehistoric burial mound set by the A4 road. This too dates from the Bronze Age, about 1500 years BC and is beautifully shaped.

AVEBURY lies off the A4, 8 miles west of Marlborough.

BATH
Avon

The city of Bath is one of the historic glories of England, and it has been so since Roman times. When the Romans built their first city here, as a spa, they called it *Aquae Sulis*, The Waters of the Sun, and Bath is still a spa, and never more beautiful than when seen from the south as the sun gòes down in the evening, for the light then casts a deep red glow over the golden Cotswold stone.

The Bath we see today is a Georgian city, built during the 18th and early 19th centuries by three architects, the brothers Wood and William Pultney, urged on by that elegant man of fashion, Beau Nash, who made Bath a centre of society and the Queen of English Spas. The present city reflects that gracious age and, for a detail, note the passages

Soft Cotswold stone and fine Georgian

architecture combine to form this attractive curving terrace of houses in Bath

and alleys off the main streets, just wide enough to permit the passage of a sedan chair. Grander sights in Bath include the magnificent sweep of The Royal Crescent, elegant Pulteney Street, or the splendid views from Pulteney Bridge, which rivals anything they have in Venice. One unique attraction in Bath is the Pump Rooms, built in 1785 and made famous by the works of Jane Austen. They are still open, close to the town centre and just the place for morning coffee or afternoon tea, taken to the tunes of a string ensemble.

Bath is a place to stroll about in, full of splendid sights and interesting museums. Do not neglect to visit the Bath Museum of Costume in the Assembly Rooms, where one exhibition is devoted to a history of knickers, or the American Museum in Britain at Claverton, four miles to the south-east. That, plus the Abbey and the Roman Baths, will take up at least three days of your time.

BATH lies on the M4 motorway, 80 miles west of London.

BODMIN MOOR
Cornwall

Cornwall is more famous for its coast than its hinterland, and there are no wild inland places here to match those of Dartmoor or Exmoor in Devon, but Bodmin Moor, which lies between Bodmin town and Launceston, and is straddled by the A30 trunk road, is a pretty, attractive alternative. Bodmin Moor owes much of its fame to Daphne du Maurier's *Jamaica Inn*, which is set on the moor, while an even more legendary spot is the Dozmary Pool, set on a bleak spot of the moor and said to be the place into which King Arthur's magic sword 'Excalibur' was thrown as the King lay dying.

Like the other Western moors, Bodmin is an open, treeless place, well scoured by the ceaseless Atlantic winds, with great views on offer from such heights as Brown Willy (1375 ft) or Rough Tor (1312 ft).

Bodmin, on the western side of the moor, is the county town, although many other Cornish towns are now much larger. It is the ideal centre for exploring Bodmin Moor and has two fascinating museums, one on the Great Western Railway Company, and the other on the history of the county's regiment, the Duke of Cornwall's Light Infantry.

BODMIN lies 22 miles west of Launceston on the A30 across Bodmin Moor.

BRIXHAM
Devon

Brixham lies at the southern end of Torbay, on that long sweep of coastline which is now known as the English Riviera, and runs south to Brixham, through Paignton from Torquay. The centrepiece of Brixham is the harbour, a sunny spot, full of fishing boats and offshore trawlers surrounded by serried ranks of pastel-washed houses.

Brixham, in spite of having a prosperous foothold in the tourist trade, is still very much a fishing port, with a considerable fleet still in residence. Another noted resident in the harbour is a replica of *The Golden Hind*, the ship in which Sir Francis Drake sailed round the world. HMS *Bellerophon* put in here in 1815, carrying Napoleon to exile in St Helena. In 1688, William of Orange landed here to start The Glorious Revolution which removed James II from the throne, and in the little church, a 19th-century vicar, Henry Lyle, wrote that seafarer's hymn *Abide with Me*. The little winding streets of

Brixham are full of evocative spots, but wherever you go you will never be far from the sight of the sea, the smell of salt and the high cries of the herring gulls. Do not leave without visiting the Fishermen's Museum, the Aquarium, and the National Coastguard Museum, all of which are close to the harbour.

BRIXHAM lies 12 miles from Torquay on the A3022.

CASTLE DROGO
Devon

The Renaissance splendour of Castle Drogo is an illusion, for this fine house was built in the present century

Visitors from abroad who travel through England soon become attracted by the blissful spectacle of the English country house. Often seen through trees, at the head of a long driveway, or across a copse-dappled parkland, they are relics of a now bygone and more gracious age, beautiful to look at, ruinous to maintain. Many have passed into the hands of the National Trust, but far too many have fallen into decay or simply disappeared.

Castle Drogo is probably the last country house that will ever be built in the old, grand manner and is therefore worth visiting. It was built between 1911 and 1930 by the famous architect Sir Edward Lutyens for his client, Julius Drew, a grocer. Even Drew's wealth could not sustain it and it passed to the National Trust in 1974.

As befits the long tradition of English country houses, Castle Drogo is a huge building, built in granite and occupying a commanding site overlooking a vast open countryside. However, as a basically modern building it also contains all the best of modern comforts, and visitors can admire the best that money could buy and good taste command, in furniture and paintings. The castle is open during the summer months only, and should not be missed on any journey through this part of England.

CASTLE DROGO is at Chagford which lies off the A382, 3 miles west of Moretonhampstead.

CERNE ABBAS
Dorset

Cerne Abbas is a pretty village, built of the most beautiful soft stone. It was once a centre for the leather trade and had a famous Benedictine abbey, parts of which can still be seen. In the village itself stands the Church of St Mary, a mainly 15th- and 16th-century building in local stone, which is one of the loveliest religious buildings in the country. This fine church was carefully restored and cleaned in the 1960s and the 13th-century church and the fine tower are now again quite splendid.

Attractive though Cerne Abbas is, visitors are drawn here not so much by any love of architecture but by the sight of that immense and rather startling hill figure, the Cerne Abbas Giant, a (very) naked figure of a man, 180 ft tall, cut out of the turf on the hillside, north of the village. The Cerne Giant bears a club and is therefore held by some to be a representation of the pagan god Hercules, and to date from the Romano-British era, AD 300-400. Others feel that it is a much earlier fertility symbol. Whatever the origins of this unique hill figure, it is still regarded as a considerable and most unusual attraction, well worth a diversion.

CERNE ABBAS is on the A37, 9 miles north of Dorchester.

CORFE CASTLE
Dorset

The Isle of Purbeck, south of Poole in Dorset, is often said by its admirers to be one of the loveliest parts of England. In the centre of this island, which is really rather more a peninsula formed by the great inlet of Poole Harbour to the north, lies the medieval town of Corfe, overlooked from the green ridge above by the spectacular ruins of a once mighty castle.

Corfe itself is a quaint, rambling little town, full of those leaning photogenic houses, but the eye is drawn inevitably to the castle walls above, which are best seen from the Arne road on the north side or from the Kingston one to the south.

William the Conqueror laid the foundations of the present castle but it was rebuilt and extended several times in the next few centuries. King John used it as a prison and it was here that he starved to death a number of captured French knights who disputed his claim to the English throne. Corfe remained a Royal castle up to the time of the Civil War, when it was held for King Charles during a siege, and only captured by treachery. After the castle fell, Cromwell ordered it to be 'slighted' or made indefensible, a fate which overtook many castles, and the walls were breached with gunpowder.

What remains is still an evocative reminder of a violent era of English history, though very beautiful to look on.

CORFE lies 5 miles east of Swanage on the A351.

COTEHELE HOUSE
Cornwall

This medieval manor house, built between 1485 and 1539, is generally regarded as the finest example of its kind in England.

It is not very large, but it is well preserved and very beautiful. From 1485 until 1947 it stayed in the hands of one family, the Mount-Edgecumbs, but they then presented it to the National Trust, ending a connection that began when Richard Edgecumb, the local lord, raised his tenants to fight for Henry Tudor against Richard III. The chapel of the manor contains a clock from this period, dating from 1489, while the house holds the results of nearly 500 years of discreet collecting, with a fine display of armour, furniture and tapestries. The park and grounds cover nearly 1000 acres, a considerable estate, and this too is beautiful, with rose gardens, ponds and a dovecote set into the natural beauty of the green Cornish landscape. Down on the banks of the River Tamar stands the manorial mill and cider press.

COTEHELE HOUSE lies at St Dominik, 2 miles west of the village of Calstock, north of Saltash.

The Cerne Abbas Giant still strides across this green Dorset hillside, high above the town

DARTMOOR
Devon

Dartmoor occupies most of Central Devon, and plays an equally prominent part in the English concourse. On this bleak and lonely moor, Conan Doyle set his *Hound of the Baskervilles*. At Princeton, in the heart of this desolation, stands Dartmoor prison, built to house captured French soldiers during the Napoleonic Wars, and still in use. When the mists come stealing in across the moors, this is a fearsome place.

When the sun shines, which is quite often, Dartmoor sparkles, a place of subtle colours, purple heather, flaring gorse, and everywhere the tinkle of streams, with sheep and Dartmoor ponies cropping the short grass. Dartmoor is the largest stretch of open country in the south of England, and is a wonderful place for walkers, campers and horse-riders, even though much of it is reserved all or part of the time for military training and exercises. It lies at well over the 1000 ft mark and is well supplied with dramatic rocky outcrops known as tors, which add much beauty and

 Dartmoor, that vast, wild expanse of rock, heather and bog

terest to this otherwise open landscape. Places to visit on the moor include Two Bridges, which lies close to Princeton in the centre, Postbridge on the B3212, Moretonhampstead, a very pleasant little town and, of course, Widecombe-in-the-Moor, east of Two Bridges, which still holds that Fair which Uncle Tom Cobbley and All were so keen to visit.

DARTMOOR lies north of the A38, between Newton Abbot and Plymouth.

DARTMOUTH
Devon

The Dart is one of the most glorious rivers in the West of England, and is never lovelier than at Dartmouth, where it broadens into a wide estuary before the sea. Here the river is overlooked and guarded by two great castles, one on either bank, which were built to defend the shipping sheltering in the green estuary which sweeps upstream almost to Totnes, crammed with sailing craft all along the way. The Royal Naval College, where officers are trained for the Royal Navy, has been at Dartmouth since 1863, and provides just one of the many links which join pretty, busy Dartmouth to England's strong maritime tradition.

Most visitors to Dartmouth find enough to do just wandering about the streets, admiring the Elizabethan houses near the quay, or the Jacobean houses along the Butterwalk, but there are boat trips to sea or to Totnes, fishing in the river or offshore, and a whole host of things to see and do here, many of them inevitably connected with sailing, for this is a great centre for West Country yachtsmen, and a popular port for visiting yachts, from across the Channel or even the broad Atlantic. Do not leave without trying a boat trip, which is the best way to view the town, or getting a good view over the harbour from the walls of Dartmouth Castle, built in the late 15th century and still in a superb state of preservation.

DARTMOUTH lies on the B3207, 30 miles east of Plymouth.

DUNSTER
Somerset

Set on the north-east side of Exmoor, little Dunster is a splendid town, full of romantic medieval echoes. The Luttrell Arms is a fine old hostelry, much restored as late as the 17th century. The famous Yarn Market is at least a hundred years earlier, and recalls the fact that little Dunster was an important centre for the cloth trade. The Church of St George, England's patron saint, contains all the true glories of an English parish church, a fine 15th-century roof, a magnificent rood-screen, brasses, flags, monuments, gleaming brass, tall vases full of flowers — all quite outstanding.

The glory of this lovely town however, is Dunster Castle, the ancient home of the Luttrell family who have lived here since 1376. Seen from the town, the castle looks like everyone's idea of a medieval fortress, but over the centuries the Luttrells have gradually changed it into a comfortable country home. They held Dunster for the King for five months in 1646, and the wall suffered severely from the ceaseless cannonades, but the interior is still intact and reflects the richness of English style from the 17th to the 19th century. The castle is now owned by the National Trust and is open to visitors on most days in summer, and like the town itself, should not be missed.

DUNSTER lies 3 miles south of Minehead, off the A396.

The Druids at Stonehenge

Dusk at Stonehenge, that unique stone circle and Druid temple

The stone circle called Stonehenge near Amesbury was erected by Bronze Age people about 4000 years ago. It is either a solar observatory or more probably a temple, and it was in that latter role that it attracted the attention of the Druids in the years before the Romans came, and their interest continues even today. Curious though it may seem, it is typical of England that a Druidic order still exists — the Ancient Order of Druid Hermetists. They assemble at Stonehenge on Midsummer's Eve and keep a vigil there throughout the night to watch the dawn break over the Sun Stone, and then hold a service on the old pagan altar where, in prehistoric times, a sacrificial victim was executed as the sun's first rays touched the stone.

Even today, mercifully without the blood letting, and in spite of the presence of too many less-than-decorous spectators, the Midsummer Druidic ceremony at Stonehenge is a unique, almost eerie spectacle. The Druids are robed in white, and wear hoods decorated with various designs. Bearing banners and sprigs of holly or mistletoe, they circle the stones to sing their hymns and listen to ancient, pre-Christian incantations.

The cult of the Druids has deep roots, which go far back into the dark ages, when it may — who really knows — have had links with the cult of Mithras, or with all those now long-forgotten religions that gave meaning to the life of prehistoric Man. Typically, it is here in England that their cult survives, revived each year.

EXETER
Devon

Although this city suffered considerably during air attacks in the early years of World War II, and rather more severely from developers in the years since, Exeter still retains a number of historic sights, and is a place with considerable charm.

The setting helps. Exeter stands on the River Exe, at the point where it begins to widen into a broad estuary of sand and mud flats which lead down to the sea at Exmouth. Setting this aside though, Exeter is an old city dating back to Roman times. The splendid cathedral, first erected in the pre-Conquest days of 1050, was rebuilt in the next century, then largely pulled down and rebuilt yet again in the 14th century, during the golden age of Gothic architecture, of which this is a magnificent example. The interior is quite glorious, with superb 13th-century vaulting and soaring columns in Purbeck marble. Outside, visitors marvel at the carvings on the West Front, while two original pieces of work are the 14th-century Bishop's Throne which slots together without the use of nails, and a very fine 15th-century clock. There is modern work here too, notably in the Chapter House.

Strolling about Exeter, especially in the parts around the cathedral, is always enjoyable, for the Cathedral Close is full of old buildings, and the High Street contains a 14th-century Guild Hall which is said to be the oldest in England — a claim disputed by some — which somehow escaped destruction during the air raids of 1940-42. Down by the quay visitors will enjoy seeing the Customs House, built in 1681, and a tour of the fascinating Maritime Museum, a superb collection of ships and yachts from the age of sail or the early days

Oare Church, in the centre of Doone Valley, by Exmoor, where Lorna was killed by evil Carver Doone

of steam, many of them afloat. This unique Museum is a must in any visit to the fine cathedral city of Exeter.

EXETER lies just off the M5 motorway, 170 miles from London.

EXMOOR
Somerset

Next to the wilds of Dartmoor, Exmoor is the largest expanse of open country in the West of England, a high, breezy place, very green, often shrouded in rain or mist. Much of the Moor lies in Somerset, abutting the border with Devon, with Exford as the central village, and plenty of little roads fanning out from there into some remote parts of this beautiful, empty landscape.

On the northern edge, close to the sea, lies Porlock, from which the 'Person from Porlock' came to interrupt Coleridge during the writing of *Kubla Khan*, which was therefore never finished. A little to the west of Porlock, a minor road off the A39 leads down to Oare and the so-called Doone Valley, the setting for R.J. Blackmore's famous romance, *Lorna Doone*. The story is fiction but the Doones really existed and were a band of 17th-century thieves, exactly as the book describes. In Oare Church, which stands on a little hill beside the road, the locals will show you the window through which the wicked Carver Doone shot Lorna as she was marrying the hero, John Ridd; R.J. Blackmore knew this church well, for his grandfather was the vicar.

The Doone Valley is pleasant farming country, steep-sided, full of little hamlets and traversed by a sparkling river, quite a delightful spot in the summer sunshine.

THE DOONE VALLEY lies 9 miles west of Minehead, south of the A39.

GLASTONBURY
Somerset

Glastonbury is much more than a town; it is a legend, and a place of pilgrimage. There is something about Glastonbury that takes the visitor back to another age, for this is a haunting, memorable spot, like nowhere else in the broad kingdom — the place of King Arthur and the Glastonbury Thorn.

To the visiting eye, the most striking part of the town is the remarkable setting at the foot of the steep Glastonbury Tor, and the fine array of old buildings, especially that ancient inn, The George and Pilgrims, which has been welcoming devout pilgrims for centuries.

Such sights apart, it is the legends that draw people to Glastonbury. Legend has it that Joseph of Aramathea came here and buried the Holy Grail, the cup Christ used at the Last Supper, under the Tor. A great abbey flourished on this story, and on the belief that King Arthur — 'the once and future King', and his Queen Guinevere are buried here. The abbey grew to great size and wealth during

Looming high above the old town and abbey, Glastonbury Tor

the Middle Ages before it was totally destroyed by the orders of Henry VIII in 1539, when the King's Commissioners hanged the last Abbot of Glastonbury at his own gate.

Legends laid aside, this is a beautiful spot, evocative of the past. Do not leave without climbing to the top of the tor, for the fine views.

GLASTONBURY lies 20 miles east of Bridgwater on the A39.

HELSTON
Cornwall

Unlike most of the places featured in this area, Helston is quite a large town. It lies on the main A394 road between Truro and Penzance, and is famous for its Floral Day celebrations which take place each year on 6 May, a relic of some pagan rite of welcome to the spring. The whole day is devoted to music and dance, and the noontide dance is particularly memorable, for the townsfolk; the men in full morning dress, pick up the tune of the Floral Dance and dance in and out of the houses.

That annual excitement apart, Helston is an interesting historic town at any time of the year. A mining centre, it has a Charter dating back to 1305 and the craft shops and stalls along Meneage Street are full of interest, as is the Monday cattle market. The Guildhall dates from 1576, and the 16th-century Angel Hotel was once the town house of the politician Sidney Godolphin. Helston is the ideal spot to stay in while touring the Lizard Peninsula. There is a bar now across Helston river, which forms a vast pool, Loe Pool, below the town, while to the north and west lies beautiful wild country, dotted with old tin mines, which should not be entered. Other sights to see hereabouts include the Seal Sanctuary at Gweek, the Camborne School of Mines Museum on the A3047 and, if you must visit a mine, you can explore the Poldark Mine at nearby Wendron, which has the old machinery in operation, as well as a bookshop and museum.

HELSTON lies 13 miles east of Penzance on the A394.

ILFRACOMBE
Devon

Ilfracombe lies on the north coast of Devon, and looks out across the Bristol Channel to the green and misty coastline of Wales. The harbour is, as always in these West Country ports, the principal place of interest, and it was once a centre for coastal cruise ships, plying across the Bristol Channel to ports around the shore. These have all gone except for the one to Lundy Island, but Ilfracombe is still a pleasant place and an excellent base for touring this attractive coast east to Lynton and Lynmouth, and up onto the wastes of Exmoor, west to the vast sands of Woolacombe in Morte Bay. There is an Iron Age fort on Hillsborough which overlooks the harbour, an interesting town museum, and a Tropical Wildlife Zoo in Bicclescombe Park. The town beaches are plentiful, if rather small, some of sand, some of shingle, but all have rock pools to paddle in, and sheltered sun-traps to doze away the day.

ILFRACOMBE is 13 miles north of Barnstaple on the A361.

LAND'S END
Cornwall

From the high breezy cliffs of Land's End, pounded by the full fetch of the Atlantic waves, the visitor to Britain looks out towards America, 3000 miles away or more. This is the most Western point in England, and although it is naturally very crowded with visitors in summer, it is here, far more than elsewhere in Britain, that you really appreciate that Britain is an island.

Those who manage to visit this spot in winter, when the westerley gales come roaring in from the Atlantic, will find the seascapes and spray quite awesome to behold. Out to sea, the tall spire of the Longships Lighthouse blinks a warning to passing ships, in spite of which they regularly come to grief here. The Islands far out on the horizon are the Scillies.

Close by is a host of attractive little villages, along the coast or tucked away just inland, like Sennan or St Just, while Penzance is just a few miles away.

LAND'S END lies 10 miles west of Penzance on the A30.

LUNDY ISLAND
Devon

Lundy Island lies well out in the Bristol Channel, off Hartland Point, and can only be reached either by helicopter or on a twice-weekly boat service from Ilfracombe or Bideford.

It now belongs to the National Trust, though the properties here have been converted into holiday homes for let by The Landmark Trust, which devotes itself to the preservation of old buildings of all kinds. About 40 people live on Lundy all year, and amenities on shore include one pub, The Marisco Tavern, a one-man brewery, a small shop and a lot of sheep. The principal residents are seabirds and it is Lundy's reputation as a centre for birdwatching that draws the visitors throughout the year.

It was once a refuge for pirates, and can still be a wild spot in winter when the great gales come booming up the Channel from the Atlantic, but in Spring and Summer Lundy is delightful, full of birdsong, mellow with sunlight, a great curtain of sea-pinks sweeping down the slopes of the western cliffs towards the blue sea far below. It is not the easiest place to get to but those who make the effort find Lundy the perfect island, and tend to return there again and again.

LUNDY lies off Hartland Point on the north coast of Devon, and is best reached from there by helicopter. The flight time from Hartland Point is 7 minutes.

LYME REGIS
Dorset

Lyme Regis is a beautiful place, very historic, very photogenic. The suffix 'Regis', 'the King's town', was granted by King Edward I, 'The Hammer of the Scots', in the middle of the 13th century, and that famous curving breakwater, 'the Cobb', which has featured in such novels as Jane Austen's *Persuasion* and John Fowles's *The French Lieutenant's Woman* was built to protect the harbour in the 14th century, and is still intact today. In later centuries Lyme Regis saw the first sea fights against the Spanish Armada (1588) and in 1685 the Duke of Monmouth landed here to start that ill-fated rebellion that led to his execution — history is everywhere in this little Dorset port.

Lyme Regis today is a very pretty spot, set in a wide bay, and the chalk cliffs hereabouts are noted for fossils, many examples of which can be seen in the town museum. Best of all though, just sit somewhere on the beach and watch the boats, or tread the path of the fictional heroines and walk out on the Cobb for a fine view back into the town.

LYME REGIS lies 25 miles west of Dorchester by the A35.

MARLBOROUGH
Wiltshire

Marlborough is an old coaching town, a way-station on the road from London to Bath. It still looks like a stage-stop, for the broad High Street which leads west to the precincts of Marlborough College, one of England's most famous public schools, is still lined with old coaching inns. If Mr Pickwick and his friends were to be seen strolling by, no one would be a bit surprised!

In fact, Marlborough pre-dates coaching days of the 18th and early 19th centuries by many hundreds of years, in spite of all that central Georgian architecture. The Bath Road it lies on was a Roman military road, there was a Norman castle in the place where the College now stands, and legend has it that King Arthur's wizard, Merlin, is buried hereabouts. A fire in the 16th century destroyed the medieval town.

MARLBOROUGH lies 16 miles east of Calne, on the A4.

The elegant facade of Marlborough College

MOUSEHOLE
Cornwall

Mousehole could only be a Cornish village — and they pronounce it 'Muzzle' locally, by the way — quaint, picturesque, busy, and looking out to sea, in this case across the sweep of Mounts Bay and the open sea towards Lizard Point. Until quite recently, Mousehole was a prosperous village, and a great centre for the pilchard fleet, but this excellent fish has now vanished from these Western waters and the village depends increasingly on tourism. Fortunately, they come in droves, for Mousehole is attractive, with pastel-painted houses, built in Cornish granite, and a pretty harbour full of small boats in a crescent-shaped cove.

Mousehole has long been an artists' centre, certainly since the 18th century, and they still flock here, drawn by that clear Atlantic light, the con-

trasting colours of the sea, the boats and the serried cottages along the quay. Apart from its own very strong attractions, Mousehole is a good centre for touring Western Cornwall, ideal for fishing trips or walks, or simply sitting about in the sun, watching the crowds go by.

MOUSEHOLE lies 3 miles south of Penzance, off the B3315.

NEWQUAY
Cornwall

Set on the north coast of Cornwall, overlooking a wide, sandy beach and a great open bay, Newquay is an elegant town, full of fine hotels, and the principal seaside resort in Cornwall. Until the end of the last century there was nothing here but a few fishermen's huts, so the growth of Newquay has matched that of holidaymaking in the West Country and has grown by leaps and bounds over the last 50 years. The town has become well established in recent years as a surfing centre, and when great waves sweep in from the Atlantic, and 'the surf is up', the sea offshore is full of weaving surfboards — a wonderful, exciting spectacle.

The coastline is a diverse mixture of cliffs, beaches and bays, but one particularly memorable sight lies five miles to the north beyond Walingale Bay at Bedruthan Steps. There is a fine sandy beach here, dominated by a line of vast rocks which culminate in a line of 200 ft-high offshore stacks. One of these greatly resembles the profile of Good Queen Bess.

NEWQUAY lies 17 miles north of Truro, on the A3075.

PLYMOUTH
Devon

Plymouth is a fine old English city, a great place to remember those fighting seamen of Elizabethan England who sailed from here to thrash the galleons of Phillip II's Spanish Armada. Those stormy times and the city's maritime connection can be best evoked up on Plymouth Hoe, where there is still a bowling green. It was here that a messenger interrupted Sir Francis Drake with the news that the Spanish Fleet had just entered the English Channel, but Drake was unperturbed. 'There is time to finish this game and beat the Spaniards too,' he said, and proceeded to do both.

From the Hoe the visitor can look across Plymouth Sound to Drake's Island, and there are sure to be one or two ships of the Home Fleet slipping in or out of Devonport, for this is still a great naval base.

As such, it was flattened by aerial bombardment during World War II, and much of what the visitor sees today, off the Hoe at least, has been constructed since the 1950s. Look about though, and there are still some historic sights to see. Around the Barbican there are old houses and narrow streets leading down to the Mayflower Steps from which the Pilgrim Fathers embarked for the New World in 1620. From the Citadel on the Hoe there are great views out to sea, to the Eddystone lighthouse and the green heights of Mountbatten, while those who want to see more of Plymouth's maritime connection may be able to visit the Naval Dockyard; Britons only in here though, I'm afraid.

PLYMOUTH lies 75 miles west of Exeter on the A38.

ST IVES
Cornwall

St Ives is the picture-postcard town of Cornwall, a necessary stop on any journey through this delightful and very different English county. Until about 100 years ago, this was a fishing port, thriving on vast catches of pilchards, but when this trade declined, the town took on a fresh lease of life as an artists' and tourists' centre. Among those who have painted scenes hereabouts were Sickert and Whistler, as well as the potter Bernard Leach, and the sculptor, Barbara Hepworth, and her house, now a museum, is one of the sights of the town. It is not hard to see why artists flock here, for the town is nothing if not picturesque, all narrow cobbled streets, winding lanes and leaning pale-washed houses. The street names are another minor fascination and owe their origin to the constant visits of John Wesley, the father of Methodism; Virgin Street, Teetotal Street, and Salubrious Street, are just a few of them, although they all contain a pub or two today.

A day could be spent wandering about the harbour, casting an eye over the artists working at their easels. There are fine examples of local artists' work in the popular Penwith Gallery, while those who like more fresh air and walking on their holiday will find plenty of sandy beaches and sheltered coves for bathing.

ST IVES lies 13 miles north-east of Penzance, on the B3311.

ST MICHAEL'S MOUNT
Cornwall

No visit to Cornwall would be complete without a visit to the little island of St Michael's Mount, which can be reached along a causeway at low tide, or by boat. St Michael's Mount lies in Mounts Bay, just off Penzance. The island is topped by a splendid castle, which although much restored, dates back in part to the 11th century. Legend has it that St Michael's Mount was once a haven for the wandering Celtic saints, or even that it is all that remains of the lost Kingdom of Lyonesse, much written about in the legends of King Arthur and his Knights, which has long since sunk beneath the stormy waters of the Atlantic. The castle is a National Trust property, open to visitors on most days in summer and well worth a visit. The monastery chapel dates from the 14th century, and the halls now contain a fine collection of armour, paintings and furniture, while from the battlements and walkways

The castle-topped island of St Michael's Mount seen from Marazion

the visitor can get beautiful views across the bay and along the Cornish coast by Marazion.

ST MICHAEL'S MOUNT lies ¼ mile offshore near Penzance.

SALISBURY
Wiltshire

Those who know England's fine old cathedral cities well, rate Salisbury among the finest of them all. As English towns go, this is not an old city, dating only from the 13th century, but in spite of the traffic which crowds the streets today, it remains a beautiful place, with the great spire of Salisbury Cathedral as the crowning glory. Before visiting that vast and magnificent building, take time to look about the town which is full of fine architecture and pleasant shops. The Guildhall of the Shoemakers is 17th century, that of the Joiners' Guild a hundred years earlier. The Old George Inn is old indeed, and was serving ale as long ago as 1320. Salisbury is full of such places, on and over the old bridges which span the Avon, bringing the visitor past the bookshops and out onto the green by the Cathedral, where the Cathedral Close provides the most splendid setting for the great church.

The spire of Salisbury Cathedral is 424 ft high. Built quickly between 1220 and 1258, the structure is unique among English Cathedrals for the purity of the style which is entirely Early English Gothic.

SALISBURY lies 25 miles north-west of Southampton along the A36.

SHERBORNE
Dorset

The buildings of Sherborne are this little town's greatest and most enduring attraction, and what fine buildings they are. Sherborne is a medieval market town, with an ancient public school for boys dating from 1550, two castles, a fine abbey church, and a host of other buildings solidly built of stone.

The castle to the east of Sherborne was built by the Abbot, Bishop Roger in 1107. Note it carefully, for it was one of the first concentric castles ever built in England. Sir Walter Raleigh lived here, but the castle was one of those which Cromwell knocked about a bit in the Civil War, and is therefore now more of a romantic ruin than the one Sir Walter knew. He built Sherborne Castle, and this still stands.

On any day in summer, Sherborne appears as one of those dreamy little English towns, so right here, so hard to imagine anywhere else. There is a timeless air about it, and in the detail of the buildings, many signs of an historic past; a Saxon doorway; the great bell in the Church of St Mary, donated by Cardinal Wolsey; the brasses, effigies and stained glass.

Sherborne is one of the great little towns of England and must not be missed during any journey through this part of the island kingdom.

SHERBORNE lies 5 miles east of Yeovil on the A30.

STONEHENGE
Wiltshire

Stonehenge is remarkable. At first sight, seen from a car when rushing past down the main road, this famous site looks much smaller than it appears on any photograph, but if you stop and look closer, the effect is most impressive.

Excavated by the antiquarian John Aubrey in the 17th century, Stonehenge today is seen distinctly, much as it was in prehistoric times, as a circle of standing or arched stones, all of great size and weight. No one really knows what it is, or how the great cross-members were manoeuvred into place; perhaps it was a temple or an observatory. Certainly the main axis is in line with the Sun's path on 21 June, the longest day of the year, a fact which has drawn Druids and Sunworshippers in recent years to a Solstice Festival. The stones came from as far away as Wales, again no one knows how, and estimates have it that Stonehenge was erected between 1900 and 1300 BC, a considerable time span for any work.

STONEHENGE lies 3 miles west of Amesbury off the A303.

The newly restored West front of Wells Cathedral

STOURHEAD
Wiltshire

Stourhead is now a National Trust property. This house and the beautiful gardens which surround it is the pearl of their extensive collection, and draws visitors by the thousand throughout the year. The house was built in the 1720s in the Palladian style, and although it was badly damaged by fire in 1902, enough remains to provide the gardens with a splendid backdrop. The house itself contains an extensive collection of Chippendale furniture and splendid 17th- and 18th-century paintings.

The glory of Stourhead though, lies in the gardens. They were laid out by the owner of Stourhead, Henry Hoare, between 1741 and 1750, in the Italian style, with water and shrubberies interspersed with grottoes and little temples. In the late spring, when the azaleas and rhododendrons are in full bloom, the sight and scent is quite stunning, especially when viewed from across the ornamental lake at evening time. Allow several hours to explore Stourhead, and do not leave before dropping down to the village of Stourton, or into the nearby town of Mere.

STOURHEAD lies 3 miles north of Mere, off the A303.

WELLS
Somerset

Wells is the smallest cathedral city in England, but a perfect gem. It was once chiefly noted as a dairy farming centre, producing some notable cheeses, but today it is the setting and the architecture which draw the summer visitors, and as in all these ancient English cities, the visitor would be well advised to park the car and enjoy the sights quietly, on foot.

The Cathedral is so magnificent that you will need to build up to it, or all else will pall. Look at the City Arms public house, once the town gaol, and the leaning houses around the church of St Cuthbert before you go on to inspect the vast and recently restored West Front of the Cathedral.

Wells is an ancient bishopric, dating back to AD 909. The Gothic Cathedral dates from between the 12th and 14th centuries and is a glorious example of that soaring style. The great array of sculptured figures at the West Front is best viewed through binoculars as are the actions of the 14th-century astronomic clock where knights appear to tilt with each other every hour. Inside, the Cathedral is equally magnificent and interesting, with stained glass and fine vaulting. After an hour or so here, cross the green to the Bishop's Palace, where the swans floating on the moat are wise enough to ring a bell at mealtimes. Wells deserves a day or two's stay.

WELLS is 5 miles north of Glastonbury on the A39.

Hampton Court
R.H.S. Gardens Wisley
Guildford
Abinger
SURREY
M 3
A 3
HAMPSHIRE
Winchester
Romsey
The New Forest
Bucklers Hard
The Solent
The Isle Of Wight
Portsmouth
M 27
A 27
Chichester
Midhurst
A 29
WEST SUSSEX
Arundel
Brighton
A 27
South Downs
Alfriston
Beachy Head
The Weald
M 23
M 25
Hever Castle
Penshurst Place
Tunbridge Wells
A 21
EAST SUSSEX
Bodiam Castle
Rye
A 259
New Romney
KENT
Rochester
M 20
M 2
Canterbury
Leeds Castle
A 2
The North Downs Way
M 20
Dover
Sissinghurst Castle
ENGLISH CHANNEL
0 10 20 30 km
0 10 20 miles

Southern England

A shimmering stretch of the River Wey as it wends its way through Guildford in Surrey

ALTHOUGH THE TITLE is usually reserved for the orchard crowded parts of Kent, the term 'The Garden of England' could just as easily be applied to all the counties which make up the south-east corner of Britain.

This is the part of England which civilised man has inhabited in an unbroken chain for well over 2000 years, and the mark of man has been almost invariably benign. Even the New Forest owes its state and existence to William the Conqueror, and those who travel across the South country will marvel continually at the constant beauty of the English garden, whether it lies outside a suburban villa, beneath the walls of a castle or a country house, or in the careful cultivation of a public park. There are the great gardens of England, like those owned by the Royal Horticultural Society at Wisley, or the Savill Garden by Windsor Great Park. There are the historic gardens like those at Hever Castle, or the classic ones like those created at Sissinghurst by Vita Sackville-West. Above all though, there are the glorious, infinitely varied garden created by millions of ordinary folk.

The traveller passing across the South of England would be well advised to wander, keeping off the motorways and main roads, exploring instead along those little lanes that seam this well-tended countryside. Chalk provides the sub-strata, rearing into view along the North and South Downs, in rounded open hills, making a bed for trout streams.

The beautiful thatched, timber-framed Clergy house at Alfriston

ABINGER
Surrey

Abinger, and the adjoining hamlet of Abinger Hammer, is one of those delightful English villages. It lies in woodland, below the south slopes of the North Downs, and was once an industrial centre, noted for the smelting of iron ore by the 'hammer pound' method, which gave Abinger Hammer its name. The local iron industry flourished hereabouts until long after the Civil War, but the only traces of such industry here today are the village smithy and the village clock, where the hours are still marked by a working model of a smith striking a bell. The village green has a pair of stocks and a whipping post, and the setting, old red houses set against the green woods, is never less than delightful.

ABINGER lies 5 miles north of Dorking.

ALFRISTON
East Sussex

Alfriston is a jewel of the Downs. Once, like so many of these little villages close to the shore, it was a seaport, and a smuggling centre. The smugglers would gather to divide their spoils at The Star Inn, which still stands in the narrow High Street of the village, and is one of the oldest inns in the county. The Parish Church stands a little apart, on a small hill by the river, and is known as 'The Cathedral of the Downs', not least because it is far too large for any possible congregation. The Clergy House, thatched and half-timbered, was built in 1380, and the stone cross in the Market Place is the only one in Sussex outside Chichester, which gives some idea of how important this village once was.

ALFRISTON lies 3 miles north of Seaford by the B2108.

ARUNDEL
West Sussex

Arundel is most famous for its castle and the family which live here. The castle was built to defend a gap in the South Downs and has been a home for centuries of the Dukes of Norfolk, Earls of Surrey, and Earl Marshals of England. The best view of Arundel is from the south across the River Arun, which is still tidal here. Across the Arun the town and castle march together, a splendid sight in the clear light of a summer day.

Lyrical though one can wax about it, it has to be said that the present Arundel is not all that ancient. Cromwell's artillery smashed the place in the 1640s, and it was not rebuilt until the 1780s, and only fully restored at the end of the last century. It still looks magnificent.

The town contains many old houses and two fine churches, the parish church which dates from the 14th century, and the 19th-century Roman Catholic Church of St Philip Nevi, which contains the tombs and memorials of the Fitz-Alan Howards, Dukes of Norfolk, and England's premier Catholic family. There is a Heritage Centre in the High Street and at least one splendid inn, inevitably called the Norfolk Arms.

ARUNDEL lies 12 miles east of Chichester on the A27.

BEACHY HEAD
East Sussex

Beachy Head is a splendid spot, a great jutting headland in the chalk cliffs which march along the Saxon Shore, with great views along the South Downs to the west, out to sea, or east to Pevensey Bay, where the Conqueror landed in 1066, even to the matching headland at Dungeness. Below the sheer cliff, lies Beachy Head lighthouse, and from Birling Gap, a little to the west, a narrow road slices up into the Downs, cutting the route of the South Downs Way, which begins on Beachy Head and runs for 80 miles across Sussex. Below this headland lies the town of Eastbourne, one of the South Coast's more elegant seaside resorts.

BEACHY HEAD is 3 miles south west of Eastbourne off the B2103.

BODIAM CASTLE
East Sussex

Seen across the vast expanse of lilypads that now fill the moat, Bodiam Castle is the archetype of every sandcastle, a fine example of a 14th-century fortress, and is in an excellent state of preservation. Historical detail apart, Bodiam is a romantic place, an evocation of the Age of Chivalry.

Stay a little away though, and do not linger inside the walls, for the interior of Bodiam is no more, and those splendid walls contain an empty shell. Cromwell passed this way and his men sacked the castle before putting it to the torch. That incident apart, Bodiam, although built in 1385, never saw action. It is now being gradually restored to its original glory and rests in the secure hands of the National Trust.

BODIAM CASTLE lies off the A229, 3 miles south of Hawkhurst.

BRIGHTON
East Sussex

Brighton is Britain's most famous seaside town, a lung for London and an attractive little Regency town in its own right. It has a two piers, one in urgent need of restoration, a long

Moated Bodiam Castle is a splendid example of a 14th-century fortress

pebble beach, well covered with bathers throughout the summer, and a vast quantity of hotels along the front, and above the arches of the Marine Parade. One of the great attractions of modern Brighton is 'The Lanes', a narrow, winding network of alleyways, full of antique shops, bistros and boutiques, and part of the old fishing and farming village of Brightlingstone, which stood here before the Prince Regent and his friends made Brighton fashionable between 1780 and 1820. Their town still reeks of the Regency, with the amazing cupolas of the Royal Pavilion as a centrepiece, but today most visitors to Brighton go there for the good food, the bracing air, and the great variety of nightlife available in this lively and attractive seaside town.

It is hard to say why the Prince Regent commissioned his friend, the architect John Nash, to design and construct the Royal Pavilion. Even today, when we are used to the unusual, it looks odd and not a little out of place; what people must have thought in 1820 when this onion-domed pleasure palace was finally complete, hardly bears thinking about, but it appears that opinions were sharply divided. Queen Victoria, surprisingly enough, liked it, although she transported its furniture and treasures to Windsor but most arbiters of taste, then and since, have laughed at it — and why not — for the Royal Pavilion was built for the Prince Regent's amusement. The Prince loved the fantasy world of the Thousand Nights, and this Arabian, or Chinese, or purely baroque extravanganza, has been amusing visitors ever since, although the poor Prince Regent died in 1827, and only enjoyed it for a very short time.

BRIGHTON lies 60 miles from London by the A23.

BUCKLERS HARD
Hampshire

Bucklers Hard may not look up to it today, but in the great days of sail, many of the 'Wooden Walls of England', the ships of Nelson's navy, slid into the water here. It lies on the tidal waters of the River Beaulieu, and among the famous ships launched here was HMS *Agamemnon*, Nelson's favourite flagship.

The village street slopes down to the shore between widespread 18th-century houses, and at the top stands the Maritime Museum, full of information on ships and the sea, with diagrams and models illustrating the former history of Bucklers Hard. Not far away lies Beaulieu Abbey, a popular tourist centre and home of the Montagues, which has a very popular motor museum, a monorail and park.

BUCKLERS HARD lies 10 miles south of Lyndhurst on the B3058.

CANTERBURY
Kent

Canterbury is a magnificent town, certainly one of the finest cathedral cities in England, but if one must choose a place to focus on, then it must be the great cathedral, scene of the famous martyrdom of Thomas Becket, burial place of Edward of Woodstock, called the Black Prince, home church of the Anglican Community, premier see of the Church of England — quite magnificent.

The first Christian Cathedral was built here as long ago as 597 by St Augustine, but the present Cathedral was built under the guidance of a Burgundian master mason, William of Sens. Work began soon after the Conquest and was almost complete by 1170 when four knights of Henry II's household burst into the nave and slaughtered Archbishop Thomas before the High Altar. The Canterbury Pilgrimage began within months, and though much reduced, still continues, made most famous in the 14th century in the works of Chaucer.

The town is still walled, full of fine buildings worth exploring, but days can be spent in the Cathedral alone, seeing the tomb of Henry IV, or that of the Black Prince, attending Evensong, or just marvelling at the beauty of this great church.

CANTERBURY lies 16 miles north-west of Dover on the A2.

The rich colours of a modern tapestry inside Chichester Cathedral

CHICHESTER
West Sussex

Chichester is one of those towns which is very old everywhere, but older in parts. If the overall impression is uniformly attractive, this is because the town is basically Georgian, and built for the most part in that pleasant and comfortable style. On the other hand, there was a Roman camp here — any town with 'Chester' as a suffix has Roman roots — and a fine Roman villa can be viewed nearby at Fishbourne. The Saxons called it Cissa's Ceaster (Cissa's Camp). The Normans built the cathedral and the walls were built up over the centuries from around AD 250. The centre of the town is marked by a well-preserved 15th-century market cross, from which four main streets fan out across the city.

Chichester Cathedral is the only one in Sussex, and the only one in Britain with a separate 'madelaine' or bell tower. The spire was one of the shore marks for sailors making their careful way round the sandbanks and shoals of Chichester Harbour, that vast estuary that lies just to the south, and the church, now well restored, is full of fine memorials and tombs. As a place to visit, or a touring centre for the rich variety of the surrounding coast and countryside, Chichester today is hard to beat.

CHICHESTER lies 15 miles east of Portsmouth on the A27 road.

DOVER
Kent

Dover is the gateway to England, a rather busy, traffic-jammed town, set in a hollow of the famous white cliffs. A traveller would need a lot of imagination to visualise much of romance in Dover today, unless he or she were wise enough to walk up, out of the town, to the great castle which stands high on the cliffs above.

The castle was built by the Normans during the 12th century, but it stands on Roman foundations and a Roman lighthouse remains within the walls. The castle was garrisoned by troops during World War II, was defended by Hubert de Burgh in the 13th century when the French ravaged the lands round about, and it has seen plenty of action down the centuries. The oldest part is Peverell's Tower, but all the walls are intact, the keep still stands, and the whole castle is well worth a visit, both for itself and for the magnificent views across the Channel to the distant land of France, which can be clearly seen from the ramparts.

DOVER lies 15 miles south of Canterbury on the A2.

GUILDFORD
Surrey

Guildford is the capital of the county of Surrey, and it looks it today, still an old city but with two recent embellishments such as a capital city should have: the new cathedral on the hill, and the ever-expanding buildings of the University of Surrey, just to the south.

The town dates back to pre-Conquest days, and is listed in Domesday Book, but it really grew up in medieval times as a staging point on the road to the south coast ports. Most of the castle has gone, though the grounds remain as a public park, but the chief attraction of the town is the steep High Street, marked at the centre by the great clock jutting from the belfry of the Guildhall. Observant visitors, walking down here, will note the historic 'Angel Inn', a coaching centre complete with gallery and cobbled yard, the Abbot's Hospital almshouses, founded in 1619, and Guildford House, which dates from 1660 and is a fine example of Restoration architecture. Guildford today is a busy, traffic-crammed town, but those who park the car and walk about with their eyes open will find a great deal of interest.

GUILDFORD lies 32 miles southwest of London on the A3.

HEVER CASTLE
Kent

History oozes from the stones of Hever Castle, but that apart, it's beautiful, an enchanted and enchanting place. It stands on the banks of the Eden, a winding little river, and was once a proper fortress with battlements and towers. In 1462 the castle came into the hands of the Bullens — or Boleyns — and one of their descendants, Lady Anne Boleyn, was married to Henry VIII and became the mother of Elizabeth I of England, before the King had her executed. The King courted Anne in the grounds of Hever, and it remains a romantic spot, much improved by the Astors, who bought Hever in 1903 and lived there until quite recently. The castle can now be visited, and the gardens are quite delightful, the whole area evocative of the more gracious — if sometimes more dangerous — age of the Tudors.

HEVER CASTLE lies 3 miles south of the village of Edenbridge, off the B2026.

Anne Boleyn, second wife of Henry VIII, was courted here at Hever Castle

The Isle of Wight

A short voyage across the yacht-crowded waters of the Solent brings the traveller to the Isle of Wight and a different, if very English world. The Isle of Wight is a little place, just 23 miles long by 13 miles wide, but that small area packs in a great deal of beauty and charm.

The coastline, and most particularly the Solent, are the major attractions for visitors, but the hinterland lying behind the coast is very beautiful, rolling, intensely farmed, and full of little villages. Carisbrooke, a mile south of Newport, is dominated by a great castle where Charles I was held prisoner before his trial and execution, and there are marvellous views from the castle walls.

Back on the coast, most visitors begin with a visit to famous Cowes, which is divided into East and West by the River Medina. Cowes is the home of yachting, the base for the Royal Yacht Squadron, one of the world's most exclusive clubs. There are yachts here all the year round, but the place really comes alive during Cowes Week in August when, so it is said, you can walk across the Solent, stepping from boat to boat!

To the west of Cowes lies Alum Bay, a geological marvel with a multi-coloured sandstone strata — green yellow, grey and brown — set among the chalk. Here, too, at the very tip of the island, stands the Needles, tall fingers of rock projecting from the sea. To the east lies the Nab Tower, one of a series of sea defenses built to project the Solent and the Naval base in Portsmouth Harbour.

Although the coastline is beautiful and studded with fine towns, there is history too. This was a Roman base called Vectis, and it was here, at Osborne House near Cowes, that Queen Victoria, the 'grandmother of the Royal houses of Europe', died in 1901. To those who take the trouble to go there and look about, the Isle of Wight can be a constant fascination.

Carisbrooke Castle on the Isle of Wight was once a prison

Osborne House was a favourite residence of Queen Victoria

LEEDS CASTLE
Kent

This splendid pile, set in a lake, beautifully preserved, is said with good reason to be the finest castle in all England. It straddles three islands and has a very long history, for the first fortress on the site is attributed to Ethelbert, King of Kent, as long ago as AD 857! Another, later fortress belonged to Edward I, who gave it to his wife, Eleanor of Castile, and Henry V who gave it to *his* wife, Catherine de Valois — the name 'Leeds' is a corruption of the name 'Ladies Castle'. Other ladies who lived here included Catherine of Aragon and, as a prisoner, Elizabeth Tudor.

The present castle was built by that mighty monarch, Henry VIII, who transformed the grim medieval fortress into a palace fit for a King. The Banqueting Hall is vast, the Chapel mangificent, the gardens beautiful — but the setting is superb. Swans and waterfowl cruise the moat, and the castle is surrounded by 400 acres of deer park and gardens. The castle serves as a Conference Centre, but is open on most days in Summer.

LEEDS CASTLE lies off the A20, near Maidstone.

MIDHURST
West Sussex

Midhurst is one of those places better known locally than anywhere outside a 20 mile radius, a place that people stumble upon, although few places can muster so much mellow English charm. The Angel Inn and the 15th-century Spread Eagle Hotel are just two of this little market town's attractive pubs, and the whole place is a picture, with leaning medieval houses and a number of narrow flower-fringed alleyways. The ruins of Cowdray Park, a house built by

Shakespeare's patron, the Earl of Southampton, stands just on the northern outskirts. It dates from 1520, was burned down in 1793, and is one of those places where Queen Elizabeth I definitely slept during the Royal Progress of 1591, three years after the Armada.

The Cowdray Museum is worth a visit, and the polo matches at Cowdray are still popular events.

MIDHURST lies 12 miles north of Chichester on the A286.

THE NEW FOREST
Hampshire

It is typical of England that the 'New' Forest should be 1000 years old, and was not 'new' in any accepted sense of the word, when the Conqueror came to claim it and put the animals and people who lived there under the fierce rule of the Forest Laws; woe to any man who harmed the King's deer, 'for he loved them as if he were their father'. Hunting was the great recreation of the medieval monarch, and this New Forest, large as it is, is but a fragment of those vast hunting demesnes which once covered much of England. Lyndhurst is the centre of the present forest, and visitors here can see the famous New Forest ponies cropping the grass by the roadside, visit the new Butterfly Museum, or enjoy a drink in one of the old inns. Near Minstead, a stone pillar set in the woods marks the spot where the second Norman king, the unpopular William Rufus, was killed by an arrow while hunting. The New Forest still covers more than 100,000 acres, and is still a beautiful place, and a paradise for wildlife. Even today, deer lurk in the thickets, but the sound of the King's Hunt has long since passed away.

LYNDHURST lies 7 miles west of Southampton, on the A35.

Deer and ponies still roam the dense, green thickets of the New Forest in Hampshire

THE NORTH DOWNS WAY
Surrey

The North Downs Way is one of England's oldest footpaths, running along the old track that led pilgrims from Winchester to Canterbury, and at the same time one of the country's newest designated long-distance footpaths, a fine, long, rural ramble that is never better than here, along the crest of the North Downs, between Box Hill, that smooth chalk promontory, and Westerham to the east.

From Box Hill, which lies a little to the north and east of Dorking, the visitor can enjoy vast views across the flat headland of Surrey, across the plain where aircraft rise and descend over Gatwick, even as far as those southern neighbours, the South Downs, which lie in Sussex by the sea. These Downs are chalk, but well wooded over the short grass, and although people ski down the face of Box Hill in winter, most of the North Downs Way threads a path through shady woods, down into busy little towns, like Reigate or Redhill, picking its way carefully through suburbia, as it heads east into Kent.

BOX HILL lies 2 miles north of Dorking off the A24.

Penshurst is one of the finest country houses in England, the birthplace of Sir Philip Sidney the Elizabethan poet

ORCHARDS AND OAST HOUSES
Kent

Kent has often been called 'The Garden of England' and it is certainly a beautiful and fruitful county, noted for orchards and the growing of hops. The orchards have changed a lot in recent years and are now smaller and less varied than they used to be, but they still present a glorious picture in the late Spring, when they float on a great white sea of blossom, which flows for mile upon mile across the countryside.

In other parts the tall poles which support the hop-vines stand in serried ranks; it is these hops which give bite to the good English ale. In these hopfields stand curious houses, white walled and beautiful, with unusual pointed, sloping roofs. These are the oast houses, where the hops are dried before the malting, and those funnel-roofs permit the air to circulate below, whatever the direction of the wind, apart from adding a rare and attractive touch in this beautiful garden of England.

ORCHARDS and hopfields with oast houses can be found in all parts of Kent, but especially in the Weald, south of the M20.

PENSHURST PLACE
Kent

Those who may never have heard of Penshurst will call it to mind more readily as the birthplace of that 'very parfit' Knight, Sir Philip Sidney, the Elizabethan soldier poet who died most gallantly in the Battle of Zutphen in 1586. His father, William Sidney, was the first lord of Penshurst, and the Sidney family have lived there ever since. The present owner, the Viscount de l'Isle, won the V.C. during World War II, and is clearly a chip off the great old block.

Penshurst still retains most of the agreeable medieval features, including the Baron's Hall, which has great beams and a Minstrels' Gallery, while in the dining room, the table is permanently set to display a magnificent collection of Rockingham china. In the crypt the family maintains a fine collection of armour and weapons, while other attractions in-

It is still the Sidney family home

clude a toy museum, and the beautiful Tudor gardens.

PENSHURST lies 3 miles west of Tonbridge, on the B2176.

PORTSMOUTH
Hampshire

Portsmouth has always been a Naval town, one of the bases for the British Home Fleet, and proud of it. Those who sail out of here on a cross-Channel ferry do so through lines of warships, while in the centre of the dockyard the flagship of Lord Nelson lies at rest, though still in commission, an office for the local Admiral. H.M.S. *Victory* looks much as she did when Lord Nelson sailed her into battle off Cape Trafalgar. The spot where he fell, struck down by a marksman from the fighting tops of the French Warship *Redoutable*, is marked by a brass plaque, and the place where he died, below decks, is now a Naval shrine. His most famous memorial lies all about though, in the fine ships and stout traditions of the Royal Navy, whose reputation he established on secure and still enduring foundations.

Porstmouth town was severely damaged by bombing in World War II, but visitors can see Charles Dickens' birthplace in Commercial Road, The Royal Marines Museum at Eastney Barracks, or the D-Day Museum at Southsea, as well as the Sally-Port and the remains of the old fortifications.

PORTSMOUTH lies on the A3, 17 miles south of Petersfield.

ROCHESTER
Kent

Dickens is the author who put Rochester on the map, and the citizens repay his effort today by dressing up in period costume for their annual one-week Dickens Festival which takes place in the week after the Spring Bank Holiday. The town features in many of his books under various names and guises, in *The Pickwick Papers*, *Edwin Drood*, and *Great Expectations*, and Eastgate House in the High Street is now a museum full of Dickens' memorabilia. That apart, Rochester is not the most attractive of towns, at least not visually, but it does have a number of fascinating corners. Rochester Castle, built to guard the

Nelson's flagship at Trafalgar, H.M.S. *Victory,* in Portsmouth dockyard

Medway, is a magnificent example of a square towered Norman fortress, and dates from the reign of Henry I in the early years of the 12th century. The Cathedral Church is also magnificent, and the Cathedral Library contains a fine collection of very early manuscripts and books. Put all this together, and a day exploring Rochester and the Medway makes the perfect day out from London.

ROCHESTER lies 25 miles east of London on the A2.

THE ROMNEY, HYTHE AND DYMCHURCH RAILWAY
Kent

There is something quintessentially British about the Romney, Hythe and Dymchurch Railway. It looks like a giant's toy, but it is in fact a busy, popular and useful little railway, the only link across the Romney Marsh to the point of Dungeness. It runs for a total length of just over 13 miles, and travellers can get on or alight at a number of little stations along the way. The line is open throughout the summer months, when it provides the perfect way to see the hidden sights of Romney Marsh, and engines can be inspected at other times in their sheds at New Romney.

NEW ROMNEY lies on the A259, south-east of Folkestone.

ROMSEY
Hampshire

Romsey owes its existence to the founding of Romsey Abbey, and grew up around it down the centuries into the pleasant country town we see today. The foundations of the old Saxon abbey, circa AD 1000, can be seen through a trapdoor in the nave, but the present building is entirely 12th century, a magnificent example of Norman craft and energy, but one which retains a number of relics or touches from the Anglo Saxon period. The cross, or rood, behind the altar is Saxon, as is a Crucifixion in the South Aisle. Nearby lies the grave of Lord Louis Mountbatten of Burma, who lived at Broadlands nearby, and was buried here after being murdered by the I.R.A. in 1979.

Broadlands itself is one of the finest examples of a Palladian House in England, once the home of the Palmerston family and still the home of the Mountbattens. It can be visited on most days in summer. The town of Romsey is also very pleasant, full of fine architecture and small, specialist shops.

ROMSEY lies 11 miles south-west of Winchester on the A31.

Romsey Abbey was founded by the Saxons in about 1000 AD, but the abbey-church is Norman

ROYAL HORTICULTURAL SOCIETY GARDENS
Surrey

The English are a nation of gardeners. Visitors from abroad never cease to marvel at the sight of so many flower-filled plots, and the poet who wrote that 'all England is a garden' was simply offering a well-phrased statement of the obvious. There are gardens in England of every shape and size, but if any one place can claim much of the credit for the *quality* of English gardens, it must be the Royal Horticultural Society's gardens at Wisley. Here, many new varieties of fruit, flower and vegetable are impartially tested, while established varities are improved and cross-bred. The alpine garden and the rock garden are quite outstanding, but apart from being a useful and indeed, a fascinating place, Wisley is beautiful, full of glorious flowers throughout the year, an essential stop on any journey through England for lovers or admirers of the English garden.

THE R.H.S. GARDENS lie on the A3 near Ripley, 7 miles north of Guildford.

RYE
East Sussex

Rye is one of the Cinque Ports, a group of seven — not five — towns, which held favoured charters from the Kings of England in return for supplying ships for the Royal Navy in times of peril, and transport for the

The Rock Garden at Wisley, studded with pools and planted with many rare species, is a colourful spectacle

Kings' armies during any campaigns in France. From this arose a small string of ancient South Coast towns, of which Rye is perhaps the most attractive and surprising.

The first surprise is that Rye does not actually lie on the sea; not any more. That withdrew centuries ago, so present-day Rye lies on a hill two miles from Rye Bay, though linked to the sea by a tidal channel. It is a beautiful little town, full of steep streets. The Mermaid Inn was first built in 1330, but a French raiding party destroyed Rye in 1332, and the inn was not rebuilt until 1420. The church of St Mary is very large and very magnificent, often called the Church of Earl Sussex. The nearby Ypres Tower, which contains the Town Museum, dates from 1250, when it was part of an Augustinian priory. There is a lot to see in Rye, but most people will be happy just to gaze at the views across the flats to the sparkling sea, or climb up cobbled Mermaid Street, taking photographs on the way.

RYE lies 12 miles east of Hastings, on the A259.

THE SAVILL GARDEN
Surrey

The Savill Garden is part of Windsor Great Park and lies half-hidden at the end of a lane in Egham, the village which lies on the hill above Runnymede, on the South Bank of the River Thames. Like the R.H.S. Gardens at Wisley, Savill Garden is devoted to the cultivation and improvement of rare plant species, but also produces bulbs and seeds for the Royal Parks and gardens. The water garden, the trees and flowering shrubs here are a glorious sight, especially in the early spring, and like the R.H.S. Gardens, the Savill Garden should not be missed by any garden-lover travelling through this party of the country.

EGHAM lies 3 miles south-west of Staines, off the A3.

SISSINGHURST CASTLE
Kent

Sissinghurst Castle is quite an old place, and in its time has served as a fortress and a prison. Thousands of French soldiers were incarcerated here during the Seven Years War, and they called it a *chateau*, as people still do. The old manor, much neglected, was in ruins by the 1930s, when it was purchased by Sir Harold Nicolson, the critic and historian, and his wife Vita Sackville-West. The Sissinghurst which compels visitors today is almost entirely their creation. Together they restored the house, while Vita created the wonderful and varied gardens, some of which are said to be the finest in England. There is a herb garden, a cottage garden, even a famous 'white' garden where all other colours are banished. The house and gardens now belong to the National Trust, and can be visited in summer.

SISSINGHURST CASTLE lies 1 mile east of the village, on the A262, north of Cranbrook.

THE SOLENT
Hampshire

The Solent, that narrow stretch of water which divides Hampshire from the Isle of Wight, is a fascinating and famous water. Henry V's fleet lay here, before sailing for France in 1415, so did the D-Day forces, 500 years later. The great trans-Atlantic liners came, and still come up here, borne on 40 ft tides to dock at Southampton, and throughout the year these waters are alive with all kinds of shipping, tankers from Fawley refinery, cross-Channel ferries for Brittany and Spain, and, all the time, all manner of yachts and sailing craft, not least during that great yachting festival, Cowes Week, one of the great sporting and social occasions of the English season. On the Hampshire shore, Lymington is typical of the little ports and yachting resorts which line the shores of the Solent, a place where the water groans under the keels of a thousand craft, the pubs are alive with sailing talk, and the main sound of a summer evening is the musical clatter of halyards against metal masts.

LYMINGTON lies 10 miles south of Lyndhurst on the A337.

THE SOUTH DOWNS
Sussex

Ask any lover of the South Country to list his or her favourite areas, and the South Downs will surely fall in the first three. Green, bare, windswept, rolling, they are the southern backdrop for Sussex, and look appealing from any viewpoint. Once they rested under the sea, and were smoothed and shaped by the last Ice Age; today they shelter a score of pretty places — Alfriston, Willingdon, with the Long Man, a famous hill figure, Ditchling Beacon, a well-known viewpoint, Steyning...... the list is almost endless. Here and there the bare crest of the Downs is marked by great copses or clumps of trees crowning the green hills like arboreal coronets, and from anywhere on the top the views across the county to the north, or down to the blue Channel to the south, are never less than superb, a summer paradise.

Although there are extensions and outcrops, the main part of the South Downs runs from west of Eastbourne to a point above South Harting in Hampshire, an 80 mile ridge of open

A typical view of the South Downs of Sussex, green, rolling, very English

The famous Pantiles of Tunbridge Wells, which visitors would follow down to the waters of the spa

grassland and sheep pasture, spanned by yet another famous and popular footpath, The South Downs Way, which begins at Eastbourne and is clearly waymarked all the way west.

THE SOUTH DOWNS lie just north of the Channel Coast between the A259 and the A27 and A283 roads.

TUNBRIDGE WELLS
Kent

Tunbridge Wells or, to give it its full title, Royal Tunbridge Wells, is one of the classic English spa towns, and therefore a very elegant place, set firmly in the style of the Regency period, around the beginning of the 19th century. Tunbridge Wells is not an old town, certainly not as English towns go, for most of the buildings date from the 18th and 19th centuries, especially those which line the famous Pantiles, an elegant, Italianate collonade of shops and houses. It takes its name from the fact that it was once a tiled walkway leading visitors to the mineral springs, and the waters, chanced upon by Lord North in the early 17th century, can still be drunk here. To modern visitors, Tunbridge Wells presents the air of a quiet, restful town, the sort of place favoured by retired military gentlemen.

ROYAL TUNBRIDGE WELLS lies 6 miles south of Tonbridge, on the A26.

WINCHESTER
Hampshire

Winchester is really a medieval city, and was once a Saxon capital, the focal point of the kingdom before the Conquest, and a departure point in later years for travellers on the Old Pilgrim Road to the Shrine of St Thomas at Canterbury. They left from the old church at St Cross, just south of the city, and at St Cross pilgrims can still receive a 'dole' of ale and bread before departing.

The centre of Winchester is dominated by a huge statue of King Alfred, the great monarch of the Saxons, but the town abounds in fine sights. The Norman castle has mostly been destroyed, although the Great Hall where Sir Walter Raleigh was tried, still stands. The High Street is medieval, the Town Museum, St John the Baptist's Church, and the Museum of the Royal Greenjackets, one of the great English infantry regiments, is well worth seeing. The top half of the High Street has been pedestrianised, and contains some fine shops, and plenty of cafés and pubs. For a good view over the town, climb St Giles Hill, a public park which offers broad vistas over the attractive and ancient Hampshire city.

Winchester Cathedral is one of the glories of the South, a large, magnificent church, beloved by the townspeople, still filled each Sunday by a considerable congregation. The present building dates from the post-Conquest years; it was started in 1079 and completed in 1093, although this basic structure was improved and embellished later on in the 14th century by William of Wykeham, who also founded the famous boys school which lies just across the green. The Cathedral is best remembered for the Shrine of St Swithin, that rainy-day saint, but the Cathedral cloisters, nave and garth are full of memorials to famous men and women, Earl Wavell, Izaak Walton, author of *The Compleat Angler* and of course, Jane Austen, who lived in Winchester for much of her life. The present building is full of interest, with much Norman work, a marvellous font in Tournai marble, and a host of battle flags, and brasses glowing against the old white-stone walls.

WINCHESTER lies 12 miles north-east of Southampton on the A33.

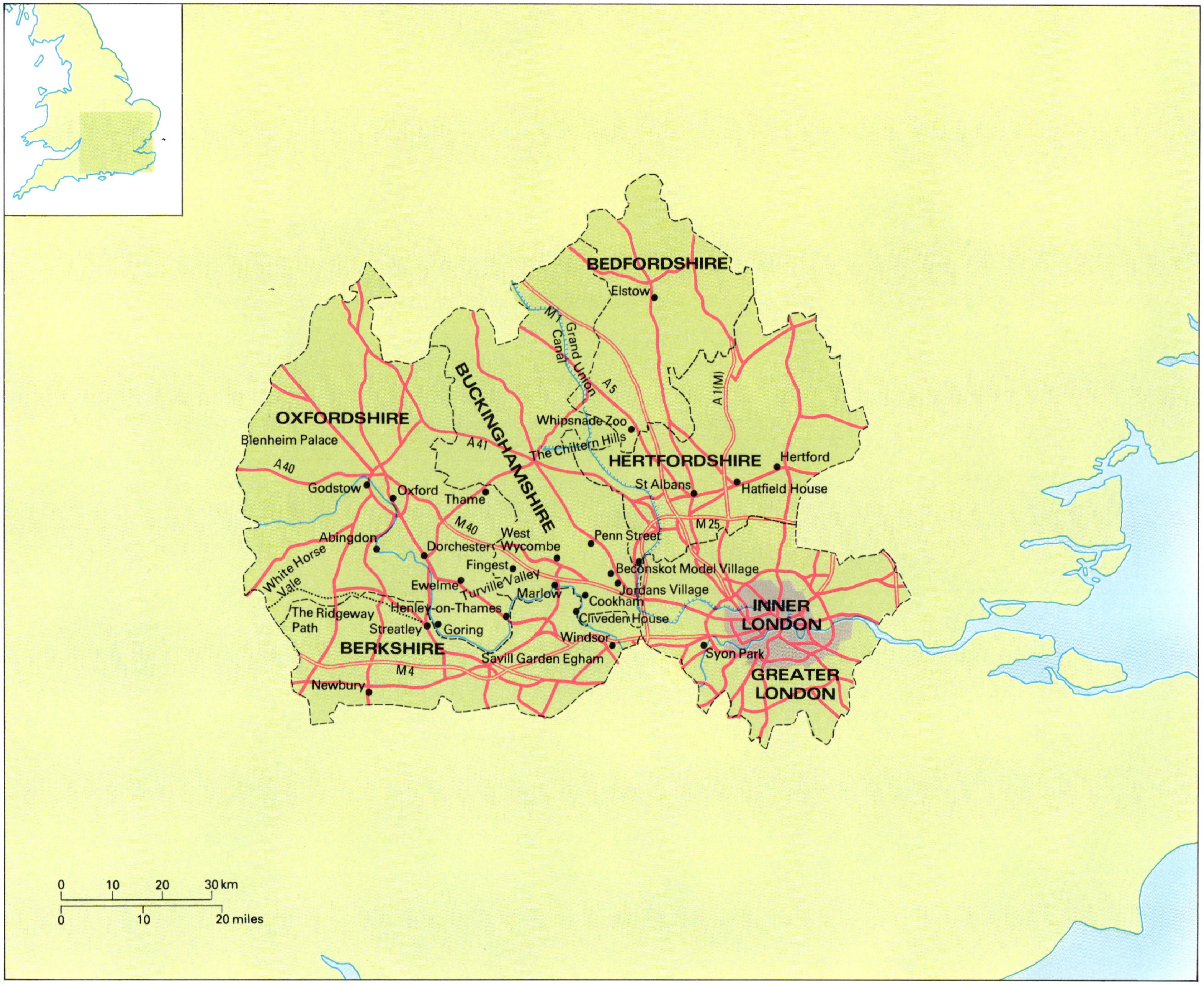
BEDFORDSHIRE
Elstow
M1
Grand Union Canal
A5
A1(M)
BUCKINGHAMSHIRE
OXFORDSHIRE
Blenheim Palace
Whipsnade Zoo
A41
The Chiltern Hills
HERTFORDSHIRE
Hertford
A40
St Albans
Hatfield House
Godstow
Oxford
Thame
M25
M40
West Wycombe
Penn Street
Abingdon
Dorchester
Fingest
Bekonskot Model Village
White Horse Vale
Ewelme
Turville Valley
Jordans Village
Marlow
Cookham
INNER LONDON
The Ridgeway Path
Henley-on-Thames
Cliveden House
Streatley
Goring
Windsor
BERKSHIRE
Syon Park
Savill Garden Egham
M4
GREATER LONDON
Newbury
0 10 20 30 km
0 10 20 miles

London, Thames & Chilterns

The bright lights of London's Piccadilly Circus in the heart of theatreland

ENGLAND IS A country of contrasts. Few regions illustrate this fact more clearly than this one, which is composed of London, the country's ancient capital, and two distinct and diverse tracts of country which lie a short distance to the west, the long and winding valley of the River Thames, which leads up to Oxford, and a little to the north, the hidden valleys of the beautiful Chiltern Hills.

The Romans built London, choosing to establish a city at the first fordable point on the tidal reaches of the Thames, the spot still marked by the long arch of London Bridge. Subsequent conquerors and colonisers, Danes, Normans, the Plantagenet kings, all found in London a worthy site for their capital. William the Conqueror began the Tower of London to overawe the City, and from that first beginning the present London grew, spreading along the then green Thames to Westminster.

One of this city's ancient jewels is the River Thames, but here in London the Thames is tidal, busy, even commercial, although now so clean and free from pollution that salmon swim upstream to the end of the tidal river at Teddington, where the Thames Valley begins. Visitors can follow this valley all the way upstream, past Hampton Court, a gem of Tudor architecture, to Runnymede, where President Kennedy of the U.S.A. is commemorated, and the English barons forced King John to sign the Magna Carta in 1215, to the Queen's castle at Windsor, and the traditional halls and cloisters of Eton College, that great public school.

ABINGDON
Oxfordshire

Abingdon was once a Berkshire town, but was transferred to Oxfordshire in the county boundary changes of the 1970s. Not much else has changed however, and Abingdon is still a bustling, prosperous Thames-side market town, with good shopping and some impressive public buildings. There are two splendid churches, St Nicholas which dates from the late 11th century and retains a typical Late Norman doorway, and the Parish Church of St Helen, much larger and endowed with a remarkable painted vault in the nave, dating from the 1390s. The most striking secular building is the Town Hall, built in the baroque style between 1678 and 1682. It is said, with some justification, to be the finest town hall in England.

Two other worthwhile places for any visitor's list are the Abingdon Museum and the County Hall. The Museum is actually housed in the County Hall, itself an ancient building containing a wide and varied range of arms, paintings, displays, tools and artifacts from Saxon times to the present day.

Those who have a low boredom threshold for architecture or museums will enjoy riverside walks from here, along the banks of the Thames, or car tours along the valley or up into the Berkshire Downs.

ABINGDON lies 7 miles south of Oxford.

BEKONSCOT MODEL VILLAGE
Buckinghamshire

This model village, of knee-high houses, is the chief tourist attraction of this pleasant little town. Set out in the shape of a modern village, this is Lilliput indeed. There are lakes and rivers, sailing ships and steamers, railway stations and a track with busy locomotives tugging trains about, and a hand-carved population of 1200 tiny people. The detail is quite remarkable and visitors to Britain can find a whole range of very English activities displayed in this quaint little village. It is open daily from Easter to early October.

THE MODEL VILLAGE is at Beaconsfield on the M40 motorway, 25 miles west of London.

The awesome facade of Blenheim Palace in Oxfordshire

BLENHEIM PALACE
Oxfordshire

Blenheim Palace at Woodstock is a stop on that popular tourist route that leads from Oxford to Stratford, but however crowded it may be in summer, it is not a place to miss.

It was built in the Royal manor of Woodstock, in the first quarter of the 18th century, a gift of Queen Anne and the nation to their great General, John Churchill, Duke of Marlborough, who beat the armies of Louis XIV in a series of brilliant campaigns. It is named after Blenheim, his most famous victory in 1704, and held by a quit-rent. Every year, on the anniversary of the Battle of Blenheim, the present Duke must present the Sovereign with a pennon, being the fleur-de-Lys of France. If he forgets, the property reverts to the Crown.

Blenheim is vast, covering three acres, and the park extends across the lake for another 2500 acres — a great estate indeed. It was built by Sir John Vanburgh, who was also a playwright, and took years to complete, is decorated in the baroque style, and although still lived in by the Dukes, house and park are open to visitors.

Sir Winston Churchill was born here, and a visit to his grave, in the nearby village of Bladon, usually forms part of any visit to Blenheim Palace. Do not leave this attractive spot without an hour's walk around the streets of Woodstock, to see The Bear Hotel and the Country Museum in the High Street, or buy a pair of gloves, the local speciality.

WOODSTOCK lies 10 miles north of Oxford on the A34.

THE CHILTERN HILLS
Buckinghamshire

The Chiltern Hills, though green and gentle, are best imagined as a clenched fist resting on the flat surface of the Oxford Plain, the knuckles representing the steep, wooded slopes of the escarpment, the back of the hand sloping away gently to the east. From out on the plain this long escarpment lies like a cloud across the horizon, dark, even a trifle menacing, and indeed this was once a dangerous place, full of highwaymen and outlaws.

Today it lies in the commuter country, close to London, and is a peaceful place, designated as an Area of Outstanding Natural Beauty, full of beech trees and long, dry valleys, well sprinkled with attractive villages and little towns. These hills lie within 30 miles of London, and cover a total area of some 400 square miles of farm and forest. This is a place to wander about in, perhaps by car, preferably on foot. There are nearly 2000 miles of waymarked footpaths, and the escarpment is spanned by one of England's most historic trails, the Ridgeway, which begins at Ivinghoe Beacon and runs south across the Thames to Avebury in Wiltshire, 80 miles of wonderful walking with superb views in all directions.

THE CHILTERN HILLS run from Goring in the Thames Valley to a point near Hitchin, Herts.

CLIVEDEN HOUSE & GARDEN
Buckinghamshire

Although there has been a country house on the site since the 17th century when the first Cliveden was built by George Villiers, the second Duke of Buckingham and a close friend of his frequent guest, Charles II, this present building is Victorian and owes its position in present memory to the American Astors. Between the Wars this was the centre for the Cliveden Set, who were said to favour Hitler, and certainly had considerably powerful political influence, not least in the person of Nancy Astor, a prominent personality who was Britain's first woman M.P. After the last War the building was handed over to the National Trust and then loaned until recently to Stanford University.

The house remains magnificent, a tribute to the power of the Victorian age and the wealth and good taste of the Astors. The gardens, which are open to the public, offer great sweeping views over the Thames Valley and the Chiltern Hills, and contain many interesting corners, not least the sword-shaped flowerbed where George Villiers fought and killed the husband of his mistress, the Countess of Shrewsbury, and Canning's Oak, under which the great 19th-century statesman would rest from his cares. It was here, during a play called *The Masque of Alfred*, that the anthem *Rule Britannia* was played for the first time at the end of the 18th century.

CLIVEDEN at Taplow, lies beside the B476, north of Maidenhead.

COOKHAM
Berkshire

Cookham is quite the most delightful village, a jewel beside the River Thames. See the beautiful little church by the river before heading up the High Street, past the ancient Tarry Stone that marked the start line for the village sports. One essential stop is the Stanley Spencer Gallery, which contains work by the late Sir Stanley Spencer (R.A.), who was born in Cookham and immortalised this village in his work between the two World Wars.

Opposite lies a famous pub, Bel and the Dragon, which has a fine restaurant and dates from the 15th century. This short, curving High Street is full of good, traditional pubs and restaurants, and leads out to the wide, green expanse of Cookham Moor, from which a footpath leads back to the river by the bridge. While here notice Mr Turk's boatyard, for Mr Turk occupies an unusual post, that of the Royal Swan-Upper, and is

Tall trees throw a reflection across the water gardens of Cliveden House

responsible for marking and protecting the Queen's swans on the river.

Look up from the bridge, towards the Buckinghamshire shore and you will see the towers of historic Cliveden House, once the home of the Astors and one of England's great country houses. Today it is an hotel, leased from the National Trust.

COOKHAM is 7 miles south of High Wycombe.

DORCHESTER
Oxfordshire

Dorchester is an attractive village, with a wide main street curving through the centre, and lined with thatched-roof cottages and antique shops, while The George is one of the most ancient and famous coaching inns in England.

However, the real glory of this place lies in the abbey, once the centre for a cathedral city founded by a missionary, St Birinius, who preached the gospel to the pagan tribes of the Thames Valley. His shrine still lies within the abbey, which is itself a shrine to the energy and good works of an American lady, Miss Edith Stedman. She came here on holiday, found the Abbey in ruins, and almost single-handed raised the money for its present restoration.

Inside there are medieval wall paintings, a magnificent Jesse window tracing the descent of Jesus, the shrine of St Birinius, and a fierce statue of a long-dead Crusading Knight. The churchyard is beautiful, and if you examine the Abbey fabric closely, you will find a most unusual corbel on the outer face of the tower. Here, carved into stone, the cheerful face of Miss Stedman looks down on her work, a part of this building she has done so much to preserve.

DORCHESTER lies on the A423, 10 miles east of Oxford.

A fine view of Dorchester Abbey from the gatehouse showing the well-preserved Norman tower

ELSTOW
Bedfordshire

Elstow is chiefly remembered for its associations with the preacher John Bunyan, author of *The Pilgrim's Progress*, who was born here in 1628. The cottage where he spent that part of his life when not in prison for his preaching, has long since vanished, but the Old Moot Hall where Bunyan would lead his followers in prayer is a magnificent red-brick, half-timbered Tudor building, and the ground floor once contained a market.

The Church of Sts Mary and Helen, where Bunyan was baptised, is basically Norman, although it contains much later work and many 16th-century monuments and brasses. Elstow Place, just to the south of the church, includes parts of the old Norman nunnery founded in the days of the Conquest, which held a two-day fair which became so notorious for merrymaking and general disorder that Bunyan used it as the model for 'Vanity Fair' in his *Pilgrim's Progress*, and Thackeray took it as the title for a novel.

ELSTOW is on the A418, just south of Bedford.

EWELME
Oxfordshire

Ewelme is not the easiest place to find, tucked away into a fold of the hills somewhere north-east of Wallingford, but those who take the trouble to find it will have discovered a perfect gem, a place which contains all that is best among English villages.

Beside the quiet main street, wide watercress beds ripple and sparkle in the sun. At the centre, set off by these tinkling streams, stands a fine group of buildings, all dating from the 15th century, each a marvel to behold.

There is the Church of St Mary, a glory in Gothic, built by William de la Pole, Duke of Suffolk in the 1450s. His wife Margaret was the granddaughter of Geoffrey Chaucer, author of *The Canterbury Tales*, and her tomb in the church, by that of her father Thomas, is blazon bedecked and quite splendid, as is the contemporary wooden font cover. In the churchyard note the tomb of Jerome

K. Jerome, author of *Three Men in a Boat*, and see that the 'K' stands for 'Klapka'.

A few steps down from the church lead into the courtyard of the almshouses, endowed by the Suffolks for '13 poore men', who must say a daily prayer for the Duke's soul, as they still do, while a little further away, lies another Suffolk foundation, the little brick-built grammar school, founded in 1437 and still serving as the oldest church school in England. Ewelme is a beautiful historic village, full of quiet little corners, perfect on a summer's day.

EWELME lies 5 miles south of Dorchester off the A423.

FINGEST
Buckinghamshire

There is no prettier place in England than Fingest in the spring, when the almond blossom is draped like deep pink snow on the trees in the old churchyard.

Fingest is a very old village, dating back to Saxon times and mentioned in that great account book of the Norman Conquest, *The Domesday Book*, when it was referred to as 'Tinghyrst'. The pale yellow tower of the present Norman church merely tops off a basically Saxon building, for although the first church rector was not appointed until 1217, the demesne of Tinghyrst — or Fingest — once belonged to Edward the Confessor. Fingest is a real country village, set at the centre of a steep-sided Chiltern valley. Those who come to see the church will enjoy a drink or a meal at The Chequers, where the way to your table leads through the gleaming brass and appetising aromas of the kitchen.

FINGEST lies 5 miles north-west of Marlow, off a minor country road, the B482.

Swan-Upping

Even though they are wild, every swan in England has an owner. Ever since the swan, that graceful bird which so enhances England's rivers, was first introduced into Britain some time in the 12th century, it has been regarded as a royal bird, and in the beginning all swans belonged to the monarch with fearful penalties for anyone killing or injuring one.

Elizabeth I made two exceptions to this rule and granted the right to own swans to two ancient livery companies of the City of London, the Vintners and the Dyers.

Ever since that time, on or about the first Monday in July, the ceremony of 'Swan-Upping' or marking, takes place along the River Thames to maintain the Royal prerogative and establish the ownership of every swan on the river between London Bridge and Henley. The number of cygnets caught and marked has declined in recent years, but in the course of a week's swan-upping some 500 cygnets are usually found and identified.

The procession of boats, rowing steadily upstream, is a colourful, romantic sight. First comes the Queen's Swan Master, in the royal livery of scarlet, his skiff and the one which follows flying the Royal Standard.

Then come four more skiffs, two bearing the Swan Master of the Vintners and his staff in green, two for the Dyers Company in blue. The rowers are wearing striped, vari-coloured jerseys and the whole spectacle makes the most beautiful picture. Each cygnet is caught and noted; those which belong to the monarch are left like their parents, unmarked, but those belonging to the Dyers carry a nick on one side of their bill and those belonging to the Vintners carry it on both sides. The whole Swan-Upping process lasts up to a week, providing a delightful spectacle.

The Queen's Swan Master on the Thames at Marlow

Set at the foot of Streatley Hill, the Bull Inn looks down the main street to the bridge across the Thames just below

GORING AND STREATLEY
Oxfordshire and Berkshire

These twin Thames-side villages occupy both banks of the Thames, at a part of the valley known as the Goring Gap, where the river has forced its way through the chalk escarpment and flows on east, leaving the Chiltern Hills to the north and the much more open Berkshire Downs to the south and west.

By any yardstick this is a beauty spot. The river tumbles in white water and is fringed with colourful flowerbeds, and usually filled with river craft and cruisers, and the villages are full of good country pubs, notably The Miller of Mansfield in Goring, which is a popular stop for walkers on the Ridgeway footpath. The Church of St Thomas, just across the street, is basically Norman, dating from 1125.

Across the bridge, enjoying marvellous river vistas on the way, the visitor arrives in Streatley. Here, one riverside pub-hotel, The Swan, is sure to catch the eye, while a drink at The Bull will refresh the walker before he or she sets out to climb the stiff slopes of Streatley hill, where the effort of reaching the top is rewarded by magnificent views over the Thames Valley. Other attractions nearby include Basildon House, and The Childe-Beale Wildlife Trust, close to the river.

GORING and Streatley are on the A329, 9 miles north-west of Reading.

GRAND UNION CANAL
Cassiobury Park, Watford

The Grand Union Canal is a great waterway, a relic of that period in English history known as the Industrial Revolution, but still very much in use, especially by cabin cruisers and holidaying narrowboats. It was built at the end of the 18th century to link the industrial Midlands to the port of London, and much of it lies deep in the countryside, or cuts through the industrial suburbs of our major towns.

The planned route across Hertfordshire cut through the estate of the then Earl of Essex at Cassiobury, and he only permitted the canal through his land on condition that he received both a large fee and the free erection of a splendid bridge across the water, which can still be seen and crossed by Grove Mill, in what is now a public park, full of trees, flowers and waterfalls, a favourite spot for picnics and weekend excursions in this part of the country.

WATFORD lies on the M25 motorway, north and west of London.

HAMPTON COURT
Middlesex

Few places evoke the might and grandeur of Tudor England so much as Cardinal Wolsey's great red-brick palace by the Thames at Hampton. Wolsey, a worldly and ambitious cleric, who rose to be Chancellor of England, built this palace at the height of his power in 1514-15, but it so aroused the envy of Henry VIII that Wolsey was obliged to present it to his sovereign, an act of generosity that did not prevent his fall from Royal favour a few months later. Henry enlarged the Palace still further and it was later beautified still more by Sir Christopher Wren.

Today, after the first impact of this rich red building has been absorbed, visitors will find plenty of interest in the house and grounds. There are magnificent State Rooms, a vast kitchen, an ancient vine which still produces grapes, and in the garden, a riot of flowers, a knot garden, and the maze made famous by Jerome K. Jerome in that classic River Thames tale, *Three Men in a Boat*, where to-

Heraldic supporters carved in stone line the bridge at Hampton Court

The 16th-century chapel of Hampton Court, where Wolsey preached

day's visitors can still become gloriously lost. Look out for the ghost of Jane Seymour.

HAMPTON COURT lies on the A308 road, south-west of London.

HATFIELD HOUSE
Hertfordshire

Hatfield House is one of the great houses of England, a splendid 17th-century palace still lived in by the Cecils, who are directly descended from Robert Cecil, first Earl of Salisbury, Counsellor to Queen Elizabeth I and James I. The previous palace to stand on this site was built in 1497 for yet another royal adviser, Bishop Morton, he who invented the stratagem of 'Morton's Fork' to extract high taxes from the nobility. (If you are shabby and live simply, you must have money saved; if you dress well and live extravagantly, you must have money to spare — either way, some must come to Our Lord the King.)

In 1607, Salisbury engaged the architect Robert Lyminge to rebuild the palace and he created the present exquisite building in the Elizabethan style with Jacobean overtones. The building took four years to complete and the size is still impressive, with a long facade and square towers on each corner. Inside it contains fine furniture, tapestries and paintings and a fine collection of Elizabeth I

Friday Street in the centre of Henley is full of overhanging late medieval houses

memorabilia. The gardens are beautiful and there is a park.

HATFIELD HOUSE lies off the A1000 on the outskirts of Hatfield.

HENLEY-ON-THAMES
Oxfordshire

Although it is a sizeable town, Henley enjoys the feel and calm of a rather pleasant village. Much of this is due to the town's position on the banks of the River Thames, in a leafy part of the Thames Valley, a peaceful spot which is only seriously disturbed once a year, in early July when Henley is the setting for the Royal Regatta and becomes a mecca for rowing folk the world over.

They have rowed at Henley for nearly 150 years. Indeed, the first Oxford and Cambridge Boat Race took place here in 1829, when the Cambridge crew wore the pink later made famous by Henley's Leander Club.

The first full regatta took place in 1839 and the event gained the Royal title under the patronage of Prince Albert, husband of Queen Victoria, in 1851, and has flourished ever since. Henley, during Regatta week, displays the England of a mostly bygone age; clean-cut young men, pretty girls in summer dresses and hats, old rowing Blues in blazers, lobster for lunch and strawberries for tea in a line of marquees along the river. Some people even go there to watch the rowing.

Out of the Season, Henley remains a pleasant town, with fine churches, good pubs, a number of nice tea shops and restaurants, and at any time of the year, that glorious Thames-side setting.

HENLEY-ON-THAMES lies 6 miles from Maidenhead on the A423.

HERTFORD
Hertfordshire

Little, quaint, red-brick Hertford is the county town of Hertfordshire, although it is now much smaller than many other towns in this county. It is a pleasant place to visit, although many of the old attractions have long since vanished, including no less than five splendid medieval churches. One place worth seeing is the 17th-century Quaker Meeting House, the oldest Quaker Meeting House still in use. Not much remains of Hertford Castle by the River Lea, except the gatehouse, but the central charm of this little town remains intact, and is best seen down the old side streets or in the many ancient inns, such as The Salisbury Arms, which was serving English ale as long ago as 1425.

HERTFORD lies 20 miles north of London, off the A10.

JORDANS
Buckinghamshire

Jordans village is, by any yardstick, a most curious place, a living reminder that many places in England retain strong links with some event in the distant past. On first sight, Jordans today looks like many other pleasant English villages, but then look closer.

The story of Jordans dates back to 1671 when a local farmer sold to the Society of Friends, the Quakers, a plot of land on which to build a Meeting House — the asking price was £4. The Meeting House was first built in 1681 during a time of religious persecution of the Quakers, whose chief spokesman was a local preacher, William Penn.

Jordans became a centre for the Quakers, and they still run this village and assemble in the Meeting House, a small, oak-panelled chapel, surrounded by a green, studded with small tombstones.

William Penn was buried here in 1718, under the usual plain Quaker headstone, and this spot, linked as it is with Pennsylvania, attracts thousands of Transatlantic visitors every year, who should also take time to see the Mayflower Barn further up the hill, which was built from the timbers of the ship which took the Pilgrim Fathers to America.

JORDANS lies a mile east of Beaconsfield, off the A40 road.

LONDON
Covent Garden

Until quite recently, Covent Garden was best known as a centre for the fruit and vegetable trade, a place where the narrow streets filled up with produce-laden trucks each evening, and trading went on for most of the night.

Now the market has moved and Covent Garden has been transformed into one of London's liveliest and most fascinating places, full of little shops, cafés, restaurants and boutiques, a place where street performers gather to entertain the crowds, the perfect spot to see the city people at play on a warm Summer evening.

While you are there though, look up and look about you, for the square, or *Piazza*, is an architectural gem. Much of it dates from the 17th century, when this square, in a then rural part of London, was laid out by the fourth Earl of Bedford and his famous architect, Inigo Jones. Above the modern cafés and shop fronts, Covent Garden has an elegant façade, so study closely such places as the Church of St Paul, built by Inigo Jones in 1633, or the most famous landmark of all, the Corinthian portico of The Royal Opera House, a stage for world-class performances, and a social centre for London society, who love to gather in the elegant crush bar.

COVENT GARDEN lies half-a-mile east of Trafalgar Square.

The old hall, once the centre of Covent Garden Market

LONDON
Hyde Park Corner

Hyde Park Corner is one of the focal points of this sprawling city, a busy, traffic-filled maelstrom, which offers much of interest to those with an observant eye. It lies at a junction of two famous parks, Hyde Park to the north, Green Park to the east, at the foot of Park Lane, sandwiched between the sweep of Piccadilly and the elegant shops of Knightsbridge.

As befits such a splendid setting, Hyde Park Corner contains some notable memorials, mostly of a military nature, not least Apsley House on the north side of the square.

This was built by Robert Adam in the 1770s, and later purchased by the 'Iron Duke', the Duke of Wellington, victor of Waterloo, when, because of his eminence in public affairs, it became known as No. 1, London. It is now a musuem, full of Wellington memorabilia.

That apart, there are fine memorials to the Machine Gun Corps of the Great War, and a particularly splendid one to the Royal Artillery, a magnificent, evocative and tragic creation. Visit Hyde Park Corner at around 10 o'clock on any weekday morning, and watch the Queen's Life Guard ride past to mount guard at St James's Palace, a colourful spectacle.

HYDE PARK CORNER lies half-a-mile west of Piccadilly Circus.

LONDON
The River Thames

Poets have eulogised the Thames at London for centuries, and those visitors who stand on Westminster Bridge and look up or down stream will find it easy to see why. The river sets off the city, the city puts the river in a timeless frame. Look west, upstream, to the Victorian Gothic of the Houses of Parliament, to the half-hidden roofs of Lambeth Palace on the South Bank, the London home of the Archbishop of Canterbury, or to the clean and glittering pinnacles of Westminster Abbey. Upstream lies the City where tall skyscrapers somehow fail to overawe St Paul's Cathedral, or the delicate tracery of a dozen Wren churches, all best seen from the river, perhaps when on a trip downstream, past the concrete temples of the National Theatre, past the brooding walls of the Tower, under bridge after bridge, to the elegant little suburb and great palace at Greenwich, now the home of the National Maritime Museum. This River Thames has been the Londoner's highway for centuries and there is no finer way to see the City and town, than from the deck of a cruiser, sailing down towards the sea. *WESTMINSTER BRIDGE lies by the Houses of Parliament.*

LONDON
St James's Park

St James's Park is the oldest Royal Park in London, dating from the time when Henry VIII seized the monastic lands and established this Park as the garden of his new palace of St James, which he was building for his new Queen and second wife, Anne Bolyn. Then the park was stocked with deer and other game, and visitors today still love to stand on the bridge which spans the lake to feed the pigeons or marvel at the flocks of colourful waterfowl which thrive here in the heart of the city. Even pelicans live on the lake and can be seen preening themselves in the rocks by the Horse Guards parade, a former tilt yard for

Tower Bridge from the Pool of London, a reminder that this was once a major port

The spires and pinnacles of the Whitehall/Westminster skyline seen from the bridge in St James's Park

the Palace of Whitehall. St James's Park is marked to the north by The Mall, a broad ceremonial way that leads from the Admiralty Arch to Buckingham Palace. Down this road distinguished visitors progress in carriages to the Sovereign's principal home, and every year, after the Ceremony of the Trooping of the Colour, which marks the Queen's Birthday, Her Majesty leads her Regiments of Foot Guards between cheering crowds.

At other times St James's is a quiet, restful spot, popular with Londoners at all times, but especially in Spring when the grass is carpeted with great swathes of Spring flowers.

ST JAMES'S PARK lies half-a-mile south of Piccadilly.

LONDON
Westminster Abbey

Westminster Abbey is one of the great historic shrines of the English people, the place where their kings and queens have come for their coronation these last thousand years.

It was built in the years leading up to the Norman Conquest by the last Saxon King, Edward the Confessor, and extensively rebuilt in the present Early English style by King Henry III in the 1250s. The outside, now cleared of London grime, is a glorious sight, and the inside, equally impressive but very different, is full of little corners to attract the curious. What a wealth of fame lies here! At the entrance lies the tomb of the Unknown Soldier, the prototype for similar memorials all over the world, and among a host of tombs and plaques to the famous, visitors may notice one to Major Andre, executed by the Americans for conspiring with the traitor Benedict Arnold. Poets' Corner contains the literati of the last several centuries.

As always in England, look about you and notice the detail. See the helm of Henry V dented on the field of Agincourt, notice the banners and stalls in the choir, see the cloisters and do not leave without visiting St Margaret's Church on the Green, a popular place for Society weddings.

WESTMINSTER ABBEY lies near the Thames, close to the Houses of Parliament.

MARLOW
Buckinghamshire

Marlow is a riverside town, cool, elegant, largely Georgian. It is best appreciated from the south, across the bridge that leads over the river from Bisham (pronounced Bizzam) on the Berkshire side. This fine suspension bridge was built by Teirney Clarke, who also built the much larger span that links Buda with Pest, and the southern arch frames the parish church of All Saints perfectly. From the bridge visitors can watch the waters sweeping over Marlow Weir, or look down on the gardens of the Compleat Angler Hotel, one of England's most famous hostelries. Tea on the lawns here is one of the great pleasures of a summer Sunday.

Marlow High Street is wide, full of good shops built into the ground floors of mostly Georgian buildings, so look up to see a good many fine examples of this graceful, comfortable period. At the top of the High Street, an obelisk marks the miles to Bath on the 'gout track', the way taken by stagecoaches bearing gout-stricken victims to the curative waters of Bath. West Street contains the house where the poet Shelley lived with his wife, Mary, while he wrote *The Revolt of Islam*, and she wrote the much more famous *Frankenstein*.

MARLOW lies on the A308, 9 miles west of Maidenhead.

St Helens Church at Marlow, viewed across the suspension bridge

NEWBURY
Berkshire

Newbury is one of England's more delightful market towns, sleepy enough today but once the scene of dramatic events, for two battles were fought here during the long years of the English Civil War (1642-1648), and those who leave their cars and walk about the town will find plenty of interesting sights. The Town Museum, set in the Old Cloth Hall, is a fine Jacobean building, and a reminder that this was the home of Jack Winchcombe, called 'Jack o' Newbury', who is said to have won a huge sum of money by betting that he could take wool from the sheep's back in the morning and have it made into a coat by nightfall.

Jack Winchcombe was certainly rich enough to entertain Henry VIII and Catherine of Aragon, to pay for the restoration of St Nicolas Church in the town centre, and establish an almshouse. That apart, Newbury has good pubs, and pleasant walks along the reedy River Kennet, or by the banks of the interesting Kennet and Avon Canal, a fine stroll on a summer evening. Two miles to the north lie the remains of Donnington Castle, a medieval fortress largely destroyed in the Civil War.

NEWBURY lies on the M4 motorway, 17 miles west of Reading.

OXFORD
Oxfordshire

Oxford has been described as both a 'city of dreaming spires' and 'an organised waste of time'. Visitors must make up their own minds.

The expression 'Town and Gown' is said to set out the two broad divisions in Oxford society, and the same might be said for the town itself. Part of it is modern, commercialised, even industrial, while the Oxford of students and visitors is tranquil, academic, and an attractive architectural treasure house. This side of the city is best seen on foot, by strolling along the 'Broad' or the 'High', to look almost anywhere around the Carfax where these two meet. Note the now re-carved heads of the Roman Emperors outside the Sheldonian Theatre, and pop into Blackwell's Bookshop to buy a map or a guide to the numerous and varied colleges of the University; Keble, red-

brick and very Victorian; Merton College, where the famous Mob Quad is the oldest in Oxford; St John's, which has a magnificent garden; Christchurch, founded by Cardinal Wolsey in the 16th century and said to be the grandest, overtopped by a huge bell tower; or, my personal favourite, Magdalen, down by the river, a timeless, beautiful place with quiet cloisters, smooth green lawns, a pleasing blaze of flowers against the old stone buildings, even a deer park.

If one must choose a central jewel in Oxford, it would probably be the buildings, quad, and collections of the Bodleian Library.

The buildings are masterpieces of their time and date from the Late Medieval and Jacobean period, and the best of all is generally conceded to be that pure example of the 15th-century Perpendicular which houses the Divinity School, although some put up a strong case for the Radcliffe Camera in the beautiful proportions of Radcliffe Square. There is so much to enjoy here that there may seem no need to step inside, but the collections are no less remarkable.

The Bodleian is one of the largest libraries in the world, with over three million books. One splendid room, Duke Humphrey's Library, contains the manuscript collection donated originally by Humphrey, Duke of Gloucester, youngest brother of Henry V. This library was completed in 1490, but virtually destroyed in the Reformation when the collection was dispersed. The present collection, and indeed the whole library, owes its existence to the Elizabethan Knight, Sir Thomas Bodley (1545-1613) who made the refurbishing and restocking of this library his lifetime's work. Another noted collection is the Radcliffe Science Library. Duke Humphrey's Library and part of the

Oxford's dreaming spires (*left*), seen from the air and (above) a typical college quadrangle

New Bodleian are open to the public.

Oxford has an atmosphere all its own, and is not a place to rush round in a few hours. There is a great deal to see, enjoy and appreciate, so spend a few days here and see the place properly.

OXFORD lies 40 miles south-west of London on the M40 motorway.

PENN STREET
Buckinghamshire

The village of Penn Street has two main claims to your attention, quite apart from its beauty. The first is its historic links with the family of William Penn the Quaker. This family founded the State of Pennsylvania in the United States, and were considerable landowners hereabouts, with numerous memorials in the area, at Penn Street, Penn, and the Quaker village of Jordans, two miles to the west of Chalfont St Giles.

The second, and perhaps more compelling reason to visit Penn Street, is the famous pub, The Hit and Miss (please note, not Hit *or* Miss). This stands across the road from the village green, and provides the village cricketers with ample amounts of sustenance during the long hours of the summer cricket matches. To sit out here, pint in hand, listening to the crack of bat on ball, and the rustle of clapping from the crowd, is to feel very much at home, and in England.

PENN STREET is 3 miles north west of Beaconsfield on the B474.

THE RIDGEWAY PATH
Berkshire/Wiltshire

The Ridgeway is one of Britain's oldest long-distance footpaths, and it follows an even longer, older route, for travellers have been walking along this green road since prehistoric times. North of the Thames, much of this modern creation follows the Icknield Way, an Iron Age route along the valley, under the loom of the Chiltern beeches, but heading south and west from Streatley it becomes the open, breezy Ridgeway proper, a splendid walk across the wind-tousled grass of the Downs.

The romanesque abbey-church of St Albans

If the path seems far too wide for any footpath, this is because this route was originally a drove road, down which sheep and cattle were driven to market. From the top there are good views to quaint little villages tucked below the hills in steep little valleys. The path leads past the site of old roman villas, past Wayland's Smithy, an Iron Age site, past the cooling towers of Harwell, and brings the walker after two or three days down to the end of the Path at Avebury in Wiltshire.

THE RIDGEWAY PATH runs from Berkshire to Wiltshire.

ST ALBANS
Hertfordshire

St Albans deserves much more consideration than it usually gets. This is a very old town, and it stands on the ruins of a Roman city, *Verulamium*, and part of this can still be seen in a park just outside the centre of the modern town.

Queen Bodicea of the Iceni tribe, led her warriors to destroy Verulamium in AD 61, and it did not really recover until the 4th century AD, when the first abbey church was built here to enhance the shrine of St Alban, England's proto-martyr, a Roman citizen martyred on the hill here in AD 209. This church became the most famous and important abbey in England, and it eventually produced the country's only Pope, Nicolas Brakespear, who was enthroned as Adrian IV in 1155. In subsequent centuries, St Albans saw a great deal of strife, and was the scene of two bloody battles during the Wars of the Roses in 1455 and 1461.

Today it is a crowded, busy market town, best visited on a weekday. Sights to see include the great abbey church, with the shrine of St Alban, the Roman ruins and theatre down by the little River Ver, which gave its name to the original town, the clock tower, and the City Museum on the Hatfield road, while a walk down Holywell Hill is always interesting.

ST ALBANS lies 20 miles north of London on the A5183.

SYON PARK
Middlesex

Syon House, and the great park which surrounds it, lies on the western outskirts of London, close to the old Bath Road. This is one of several great houses, built close to the Thames, on the outskirts of London Town, which date from the 18th century and were designed as palaces for the provincial nobility, in this case for the Dukes of Northumberland, during their attendance at Court for the London Season.

This conservatory at Syon House is the centrepiece to beautiful gardens

Syon House rests on the foundations of a monastery, founded in 1415, the year of the Battle of Agincourt. Though now encased in Portland stone, much of the exterior dates from this period, but the interior was designed and furnished by Robert Adam in the 1760s. The glory of Syon is the garden, laid out in the 18th century by a famous English landscape gardener, Capability Brown, and this has been enhanced and embellished down the centuries to present a most beautiful and varied scene throughout the year.

SYON HOUSE lies off the A4 road, 3 miles west of Central London. Nearest Underground Station: Brentford or Osterley.

THAME
Oxfordshire

Thame is an historic town, the central jewel in that flat Oxfordshire plain that spreads out beneath the Chiltern escarpment, and runs north and west to Oxford. It is a pleasant market town, full of pubs, coaching inns and antique shops, set on either side of an extra wide High Street. Thame has produced a good number of famous sons from both the town itself and from the pupils educated at the grammar school; not least John Hampden, that patriot who defied King Charles I in Parliament and raised a regiment of local men to serve in the Parliamentary army. Hampden was killed nearby in the Battle of Chalgrove Field in 1643. Other students who went on to leave their mark on English history were the cavalier poet Edmund Waller, John Milton, Dr Fell — he of the old nursery rhyme ('I do not like thee, Dr Fell, The reason why I cannot tell......') and Anthony Wood, an Oxford antiquarian of the 17th century.

Perhaps it is the spirit of these people which gives Thame such a graceful, timeless air; certainly it seems to have slipped a little out of the 20th century and is all the better for it. As points of interest for a stroll about the town, do not miss the grammar school, built in 1569, or the little Church of St Mary the Virgin, built in the 13th century and vast — a perfect example of the English medieval church.

THAME lies on the B4011, 10 miles south of Aylesbury.

THE TROUT PUB
Oxfordshire

The Trout at Godstow is one of those pubs that poets and visitors will rave about. It stands beside the Thames on the northern outskirts of Oxford, and was once part of the fabric of Godstow Abbey, when Henry II of England kept his most beguiling mistress, Rosamund Clifford, 'Fair Rosamund', back in the 12th century.

The Trout has been an inn for hundreds of years, and has long been a particular favourite with the University students, who walk out here in the evening or at weekends to drink a pint on the terrace by the river, and put the world to rights.

On a summer's day The Trout is a perfect picture, a long stone building with flowers in pots, flagged floors without and within, and great trout finning quietly in the clear waters of the river, feeding well on bread tossed in by the patrons.

THE TROUT at Godstow lies off the A34 road from Oxford to Stratford, on the outskirts of the city.

THE TURVILLE VALLEY
Buckinghamshire

Hidden away though it is, in one of those deep folds of the Chiltern Hills a little north of Marlow, the Turville

The squire played cards inside the golden ball on West Wycombe's church

Valley is not difficult to find. The first sight to strike the eye is Turville windmill, a beautifully restored white-painted smock-mill high on the crest of the hill. This is now a private home, not open to the public, but if it looks familiar that is because it was the inventor's home in the movie *Chitty-Chitty-Bang-Bang*. Climb up for a closer look before turning to gaze down on the valley and villages below; Fingest, Turville itself full of fine flint-stone cottages, and little Skirmett. At any time of the year this is a delightful spot, so try it in spring, when the bluebells form a deep carpet in the woods, in summer when every field and hedgerow is bursting with blossom, or in autumn when the fall colours of the beechwoods are a glory to behold, or even in the depths of winter, when the air crackles with frost and the holly berries are red.

THE TURVILLE VALLEY lies north-west of Marlow.

WEST WYCOMBE
Buckinghamshire

West Wycombe lies a little to the west of the much larger town of High Wycombe, and a good deal further away in time and beauty. It is worth noting that this entire village belongs to the National Trust, that excellent organisation which aims to preserve the best in Britain.

Architectural attractions apart, West Wycombe has some curious places and tales to attract the visitor. Note the great golden ball atop St Lawrence's Church. Inside that sphere, the rakish Sir Francis Dashwood used to play cards with his friends in the 18th century. Further up the road lie the Hellfire Caves, where, so it is said, Sir Francis and his friends held unspeakable orgies, which seems unlikely since the 'caves' are actually tunnels left when excavating chalk for a roadway.

The present home of the Dashwood family is West Wycombe park, also owned by the National Trust, and a fascinating example of an English country home. It was built by the first Sir Francis in 1700, and he was devoted to classical architecture; each facade presents a different style to view, Doric, Palladian, Greek and Italian. The gardens are quite outstanding, and the interior, decorated in the style of the 1750s, is a mixture of neo-classical and baroque. West Wycombe is a beautiful village, now spared much of the heavy traffic that used to pour west along the A40, and worth a full day's visit.

WEST WYCOMBE lies on the A40, a mile west of High Wycombe.

WHIPSNADE ZOO
Bedfordshire

Whipsnade was the first open zoo in the world, the first place where wild animals were kept in enclosures rather than cages. It is a strange sight to see wallabies hopping about on a green English hillside, or take a train ride through the White Rhino park, but the overall effect is pleasing to both animals and visitors. Whipsnade is famous for its success in breeding and conservation, and the 2000 birds and mammals who live there seem content with their lot.

The Zoo was opened in 1931 as the country home of the Royal Zoological Society, which also owns London Zoo in Regents Park. It covers 500 acres and contains, apart from the safari railway, a dolphinarium, and all kinds of deer, birds and beasts, the perfect place for a Sunday afternoon in the English countryside.

WHIPSNADE ZOO lies at Dunstable, 5 miles west of the M1 motorway, on the B4540.

WHITE HORSE VALE
Oxfordshire

No one really knows how long the great white horse has been running on White Horse Hill, but although it has been attributed to King Alfred in the 9th century it was probably there long before the Romans came to these islands, 2000 years ago, and is certainly the oldest of several white horses cut into the Southern hills. The Vale of the White Horse was once part of Berkshire, and the Horse still runs across part of the Berkshire Downs, even if this part of the country was transferred to Oxfordshire in the 1970s.

The White Horse itself is large, elongated, dramatic, cut out of the turf and so revealed in the chalk, stark against the green of the hill, and although a minor road leads up close to the Horse, the best view can be obtained from further out in the Vale, at Uffington, which lies four miles to the north. The famous Ridgeway long-distance footpath runs just to one side of the White Horse.

THE VALE of the White Horse lies to the west of Wantage.

WINDSOR CASTLE
Berkshire

Windsor Castle is a splendid sight, especially if seen from the river far below, or at evening time from the spur-road which runs into Windsor from the nearby M4 motorway.

William the Conqueror built the first 'motte-and-bailey' (hill and wall) castle here in 1067, as one of the defenses of London, and it served that purpose until the end of the Great Civil War in 1648. Much of the castle which visitors see and relish today owes its existence in the present form to the Hanovarian George III, who loved to visit Windsor, as his descen-

The great sweep of the Long Walk seen from the Staines road provides the perfect foreground to this view of Windsor Castle

dants of what is now the House of Windsor still do. Her Majesty Queen Elizabeth II still spends most weekends at Windsor and holds a house-party during Royal Ascot week.

Visitors will enjoy exploring old Windsor, seeing the daily march-past of the Guard, and visiting the magnificent St George's Chapel, a splendid example of English Gothic, which is the home of the Order of the Garter, Europe's most ancient Order of Chivalry. The helms, banners, and name plates of the Knights may be inspected in the choir, in a building full of the tombs of English Kings. Other attractions here include the famous Royal Dolls' House and, when Her Majesty is not in residence, the State Apartments.

WINDSOR lies 20 miles west of London, off the M4 motorway.

WINDSOR GREAT PARK
Berkshire

Half hidden behind the old streets and ancient buildings of this medieval town, Windsor Great Park is magnificent, a vast expanse of farmland and forest, great trees and small copses, full of deer and other wildlife, haunted by a famous ghost.

Windsor Great Park, large as it is, is but a remnant of a once great hunting forest that spanned much of Royal Berkshire, and as such it remains part of the Royal demense. Much of it is open to the public though, and there is no finer view of Windsor Castle than that from the bottom of the Long Walk, where it crosses the A308 between Windsor and Staines. Come here in Ascot Week and watch the Queen and her guests jingle past as the carriage procession makes its way down the Long Walk to the racecourse, while on any summer weekend you may find Her Majesty out riding here, early in the morning. If you should do so, smile, bow, remove your hat and say 'Good morning, Ma'am'. Her Majesty never gives autographs.

One part of the park which should not be missed by any lover of flowers is the Savill Garden, on the eastern side near Egham, begun in 1931 and now a riot of flowering shrubs, trees and plants.

As to that ghost, some claim to have seen it. Herne the Hunter poached the King's Deer and was flayed alive for it. He hunts the forest still, on cloudy, moonlit nights, his skin flying from his shoulders, a set of antlers nailed to his brow...... or so it is said.

WINDSOR GREAT PARK begins a mile south of Windsor town centre.

THE WASH
Wells-Next-The-Sea
Burnham Thorpe
Walsingham
A 148
Kings Lynn
A 140
The Norfolk Broads
Wroxham
Dereham
Swaffham
Norwich
NORFOLK
A 1065
A 11
Peterborough
CAMBRIDGESHIRE
Bungay
The Otter Trust Earsham
Little Gidding
Ely
A 10
Huntingdon
A 11
A 140
A 12
Southwold
Newmarket
Bury St Edmunds
SUFFOLK
Cambridge
Grantchester
Otley Hall
Aldeburgh
Lavenham
Long Melford
Duxford
Sudbury
Ipswich
Saffron Walden
A 12
Headingham Castle
East Bergholt
Thaxstead
Flatford Mill
A 131
M 11
ESSEX
A 12
Blackwater Estuary
Maldon
0 10 20 30 km
0 10 20 miles

East Anglia

There is always that salty tang of the sea in the air wherever you happen to be in East Anglia. The sky reflects the steely tints of the North Sea, never far away

THE NORTH SEA shaped East Anglia. A chill, cold, often hostile sea, it pounds the East Anglian coast, piles up the shingle between the groynes along the beaches, hollows out the little creeks where the yachtsmen come to moor and shelter, and creeps in twice a day to remind the people how much of their lives is dictated by the regular flooding of the main. Nine centuries ago this sea road from the North brought the Vikings, those Jutes and Angles who first raided around the coast, then came to settle here and give their names to the present counties. East Anglia is the land of the Angles; Suffolk, the South Folk; Norfolk, the North Folk — and in many of the little towns and villages, even the local dialect recalls that far-off invasion.

East Anglia is a very diverse region, and although the change is gradual, no two counties are alike. By and large, the area is low-lying and needs protection from the sea, which from time to time eludes all man's inventions and bursts in to flood across the county. Here there are vast bulbfields reminiscent of Holland, fen country around Ely and east of Norwich in the Norfolk Broads. Cambridgeshire, famous for its old cities and University town, is flat farming country, full of slow rivers, while Norfolk is a sweep of curving coastline, little ports and towns.

In Suffolk and parts of Essex the land is less striking, but no less beautiful, for here the visitor will find the Constable country, so well remembered from a host of prints and paintings.

ALDEBURGH
Suffolk

Aldeburgh is a quiet seaside town for much of the year until it bursts into life in June, during the two weeks of the Aldeburgh Festival, founded many years ago by the town's most famous resident, Sir Benjamin Brittan, who came to live here in 1948. The story for one of his most famous operas, *Peter Grimes*, was taken from a tale written by the poet George Cramble, who lived in the town and wrote a series of stories centred on the Suffolk coast. The long beach is overlooked by a Martello tower, one of hundreds built to defend England during the Napoleonic Wars. The Moot Hall by the shore dates from the 16th century, when the town was an active centre for the coastal trade.

Many of the Festival events take place in The Maltings at Snape, a few miles inland which, apart from the Festival, holds musical events and concerts throughout the year.

ALDEBURGH lies 30 miles north-east of Ipswich on the A1094.

THE BLACKWATER ESTUARY
Essex

The Blackwater is one of those slow, tidal Essex rivers, with high mudbanks, ambling east against the surge and out into the sea near Maldon. From here it widens steadily out into the North Sea, and the Blackwater offers the perfect haven for yachtsmen, with shelter from the north-easterly gales when they sweep down from Norway, and plenty of tidal waters for coasting yachts and dingheys. From Maldon sailors cruise up the Blackwater to Mersea Island, to Brightlingsea, even up the River Cole to Colchester or out towards the Wallet and the sandbanks of Gunfleet — a lovely spot to sample the cruising waters or the east coast of Britain.

THE BLACKWATER estuary runs east from Maldon to Sales Point and Mersea Island.

BUNGAY
Suffolk

Bungay has a longer continuous history than most English towns, as is indicated in the Saxon titles of Reeve and Feoffees instead of the more usual Mayor and Counsellors. Bungay Castle, which stands in ruins off the High Street, was once a mighty fortress with a high keep and a strong curtain wall, but very little now remains, even though many of the castle stones can be seen in the town's older buildings. The Butter Cross in the market dates from 1688, and please note the weather vane topped by the effigy of Black Shuck, the devil dog, or hell-hound of East Anglia. Bungay lies on the River Waveney and is an attractive town with easy access to the coastlines of Norfolk and Suffolk.

BUNGAY lies on the A143, 6 miles west of Beccles.

BURNHAM THORPE
Norfolk

Here, in this little Norfolk village, Horatio, Viscount Nelson was born. His father was the vicar here, so Horatio grew up with the sea not far away, and returned here throughout his life. As Burnham Thorpe's — if not England's — most famous son, he is commemorated in Burnham in both the church and in Nelson Hall next door. The church lectern was made from the wood of his flagship HMS *Victory* on which he died in the Battle of Trafalgar, and both the church and Hall are full of Nelson memorabilia. It seems a quiet little place to be the birthplace of someone who spent his life in battle, but it is a pretty spot, just right for the home of a sailor, returned from the sea.

BURNHAM THORPE lies 1 mile south-east of Burnham Market.

BURY ST EDMUNDS
Suffolk

Bury St Edmunds is a very fine, very old town, and the ideal centre for touring East Anglia. It takes its name from St Edmund, King of the East Angles, who was martyred here by the invading Danes in AD 869. His body was enshrined in a monastery here about AD 900, and Bury St Edmunds was a centre of pilgrimage throughout the Middle Ages. King Canute was just one of the abbey patrons. The Barons of England met here in 1214 to draw up the terms of Magna Carta, and though the abbey suffered from riot and the inevitable consequences of the Reformation, the Abbey Gate is still intact and quite magnificent. Down in the town, which is laid out in the grid-pattern of a medieval *bastide*, St Mary's Church contains the tomb of Mary Tudor, and Bury Moyses Hall on Cornhill is a fine 12th-century building, now a museum. As a place to relax in, or a touring centre for visits out to Suffolk or Norfolk, few towns could be more suitable than beautiful Bury St Edmunds.

BURY ST EDMUNDS lies 25 miles east of Cambridge on the A45.

CAMBRIDGE
Cambridgeshire

Cambridge is a beautiful city, rich in history, architecture and scholarship. The Romans were here of course, and built the first town astride the River

Punts on the 'backs' at Cambridge on the River Cam

The soaring Gothic columns and magnificent stained glass windows of Kings College Chapel, Cambridge

Cam. William the Conqueror built a castle here just after the Conquest, and some time in the next century the monks of Ely established a school which was the foundation of the present University. Since that time Cambridge has grown in both reputation and beauty, but not much in size, for it is really a small place, easy to explore on foot. Indeed it is so congested with traffic that anything else is impossible.

Visitors can wander at will around the college grounds, except during certain weeks in April and May when the colleges are closed to the public while the students sit their end of year examinations. See Clare College, or the beautiful Kings College Chapel, begun by Henry VII, or Christs College in St Andrews Street, where Milton wrote *Lycidas* under the mulberry trees. The first college, Peterhouse, was founded by the Bishop of Ely in 1281, while Corpus Christi, one of the most renowned, was endowed by the townspeople in 1352, or take a look at the more modern colleges of Robinson or Churchill, founded as recently as the 1960s. There is so much to see and do in Cambridge that a whole book would hardly suffice, but for a diversion, why not visit the Scott Polar Institute in Lensfield Road, a museum to Captain Robert F. Scott and his ill-fated expedition. Two essential acts in Cambridge are to take a drink in the Blue Boar pub, a student rendezvous for centuries and stroll down to the riverside to the 'Backs' to watch — or join — the young people punting on the river, a peaceful sight on a warm summer evening in this most delightful university city.

CAMBRIDGE lies off the M11, 15 miles north of Saffron Walden.

DEREHAM
Norfolk

Dereham was said by the novelist George Borrow to be '......the pattern for an English market town', but then he was a local lad and probably biased. That said, it is a very attractive place with much fine architecture in that attractive Georgian style, and a particularly interesting church, St Nicolas, built from the 12th to the 16th centuries in a mixture of Norman and Perpendicular styles, which works rather well. It was once a pilgrim church, for it contains the shrine of St Withburga, an obscure Saxon saint, who founded an abbey here in the 17th century. Her empty tomb now contains a spring which fills the churchyard well. William Cowper, that most English of poets, was born in Dereham and he was buried in St Nicolas Church, under a fine monument by John Flaxman.

EAST DEREHAM lies off the A47, 16 miles west of Norwich.

DUXFORD
Cambridgeshire

During World War II, Duxford aerodrome was a front line fighter airfield, home base for the famous Duxford Wing which fought off the German air force day after day during the Battle of Britain. How fitting therefore, that this airfield should now contain the aircraft exhibits of the Imperial War Museum, housing them here on the concrete standings, and in the World War I hangars. There are over 70 aircraft in the collection, from both world wars to Concorde, including Spitfires, Hurricanes, a B17 Flying Fortress, Mustang and a German Junkers 52. On a summer day, when they stand on the peri-track, their propellers sniffing at the sky, those long ago air battles seem only yesterday.

DUXFORD lies off the M11 motorway, 5 miles south of Cambridge.

ELY
Cambridgeshire

There is something romantic about Ely, something immediately evocative of the past. It might be the clear, rain-washed air of the Fens, it might be the countless rows of Georgian buildings, or more probably the soaring front of the cathedral. Whatever the cause, the atmosphere of Ely is gentle and sublime. The cathedral is, of course, both famous and magnificent. It was begun on the orders of William the Conqueror in the years after the Conquest, but not finally completed until the end of the 14th century, for the work was delayed when the tower collapsed into the nave in 1322. When completed though, the cathedral became the crowning jewel of the Fen country, with a marvellous octagonal tower and a Lady Chapel which is probably the finest in all England. Kings School is also marvellous, as is Ely Porta, the gateway to the medieval monastery and the old buildings beside the river. Those who visit Cambridge without diverting north to Ely, are missing a great deal.

ELY lies 17 miles north of Cambridge by the A10.

FLATFORD MILL
East Bergholt, Suffolk

John Constable made Flatford Mill famous, but it was always a beautiful place and it still is. Constable must have known the Stour Valley in which it stands from childhood, for his father was a miller and Flatford was just one of the mills the family owned along the river. Indeed, Constable claimed that his childhood home, among this beautiful, lush countryside, provided the scenes that made him an artist, and one of the present delights of exploring this 'Constable Country' is spotting the places he immortalised in his paintings. 'The Cornfield' can be identified along the lane from Flatford to East Bergholt, and indeed very little has changed at all since he wandered hereabouts with paint and easel. Flatford Mill is now owned by the National Trust and occupied by the Field Studies Council, who run courses here on the local wildlife and, of course, on painting.

FLATFORD MILL lies 1 mile south of East Bergholt, by a minor road.

Flatford Mill, once owned by John Constable's father, now belongs to the National Trust

GRANTCHESTER
Cambridgeshire

Rupert Brooke made Grantchester famous through his poetry, but this pretty village on the River Cam has been a popular place with the Cambridge student community for centuries, who come here along the river by punt on summer evenings during term time. The village is rather as Brooke described it, although the church clock will repay a look, as well as the Saxon brickwork.

GRANTCHESTER lies 3 miles south of Cambridge off the A10.

HEADINGHAM CASTLE
Halstead, Essex

Unlike most of England's Norman castles which were destroyed by time or battered into rubble by Cromwell, Headingham Castle is in a remarkable state of preservation. It was built in the middle of the 12th century by the De Vere's, Earls of Oxford, and they held the castle for the next 500 years, when they served successive kings as soldiers and admirals. The castle is surrounded by a moat, like all the best castles, and a bridge leads across the water to the *enciente*, and the keep which soars up for 100 ft and gives great views over the countryside. The second floor of the keep contains the Great Hall, which has a circular wooden gallery and relics of the original plaster. A splendid evocative place, Headingham Castle should not be missed by any lover of castles.

HEADINGHAM CASTLE lies 5 miles north of Halstead, off the A604.

HUNTINGDON
Cambridgeshire

Not many people know it, but Huntingdon produced one of the great men of English history, one of the founders of our modern Parliamentary system of government, Oliver Cromwell, the Lord Protector. He was born in a house in the High Street here in 1599. During the Civil War both Cromwell and King Charles held the town in turn, neither doing the place any damage, but it is its native son whom Huntingdon remembers today.

There is a Cromwell Museum in Market Square, a Norman building which was a school in Cromwell's day. Both he and Samuel Pepys, the diarist of Charles II's time, were educated here, and it contains a great many artifacts and memorabilia, including Cromwell's death mask. Not a place to be rushed by those people who have seen what remains of Cromwell's work in other parts of the country.

HUNTINGDON lies on the A141, 17 miles north-west of Cambridge.

IPSWICH
Suffolk

When visiting any English county, do not neglect to visit the county town. It did not rise to that position by accident and is sure to be full of fine sights and historic reminders.

Ipswich, the county town of East Suffolk, used to be a sea-port. It lies

This Customs House by the quay is just one example of Kings Lynn's fine architecture

at least ten miles from the North Sea, but it looks like a port, with a line of sea-captains' houses and quays along the River Orwell.

As a port it has been busy since Roman times, although it really developed under the Normans, and was an important town under the Plantagenets and Tudors. Do not miss St Mary Elms Cottages, built in 1467, or two Tudor houses, the Ancient House and Christchurch House, which now contains work by Gainsborough and Constable as well as much fine furniture from Tudor times to Victorian. That apart, Ipswich is a place to wander about, along winding medieval streets and quays, sniffing the sea breezes that blow in from the coast, or popping in and out of the town's old churches — a delightful spot.

IPSWICH lies on the A12, at the head of the Orwell estuary.

KINGS LYNN
Norfolk

However many places have to be missed on a journey through East Anglia, spare some time to visit the exquisite town of Kings Lynn. It was established as a port long before the Conquest, but it flourished under the Normans and continued to grow in wealth and beauty between the 14th and 19th centuries, until today it is a treasure house of English domestic architecture. The Guildhall, built in 1421, overlooks the Market Place and contains the town treasure, which includes charters from King Canute and King John. St George's Guildhall is the largest surviving Guildhall in England, and is now a theatre, but these are just the central gems, for all Kings Lynn is a treasure house where medieval streets curve down to old quays, lined with merchants' houses. From Lynn a narrow channel leads

The 15th-century Guildhall in the centre of Lavenham, finest of the East Anglian wool towns

out to The Wash, where visitors still search for King John's treasure. It was from here that Captain Vancouver, a local man, sailed in 1780 to explore the west coast of Canada, and Lynn, as the locals call it, remains much as he left it — a beautiful place and one of the finest towns in all England.

KINGS LYNN lies on The Wash, 11 miles north of Downham Market.

LAVENHAM
Suffolk

Although the centre of the English wool trade was the Cotswolds, the sheep of Suffolk paid their dues to the Exchequer and provided the profits which enhanced the East Anglian wool towns, like pretty medieval Lavenham. The streets wobble about the town centre, lined on either side with those leaning half-timbered 15th-century houses which are so pleasing to the eye, and lead to the market place with its 15th-century Woolstaplers Hall and the equally lovely Corpus Christi Guildhall flanked by Tudor shops, many with their original windows. Close your eyes to the traffic here in the centre and it is not hard to imagine this town in the reign of Richard II. During his reign the parish church, with its high, dramatic tower, 140 ft tall, was finally completed. It contains many fine brasses, carved stalls and excellent stained glass. To get an even deeper look at the Lavenham of days gone by, visit the museum in the Corpus Christi Guildhall, which now contains exhibits charting 700 years of the cloth trade.

LAVENHAM lies 10 miles south of Bury St Edmunds on the A1141.

LITTLE GIDDING
Cambridgeshire

Like a number of places in Cambridgeshire, Little Gidding has a long history and a strong literary connection which endures to the present day. The history goes back to 1625

Georgian pepper-pot towers crown the front of Long Melford Hall

when a religious community established itself here, founded by Nicolas Farrar, as a place of refuge from the world's strife. Even Charles I found Little Gidding restful. The literary connection accrues to T. S. Eliot, who found his inspiration for one of his four quartets, *Little Gidding* in the life and works of Farrar. Lovers of Eliot come here in great numbers to see the church and cottages, and breathe a little of that tranquil air.

LITTLE GIDDING lies on the B660, 5 miles north-west of Huntingdon.

LONG MELFORD
Suffolk

Like Lavenham a little to the east, Long Melford is a wool town and therefore beautiful, not least along the splendid straggling high street which is lined with antique shops and Georgian houses, leading to the green behind which stands the 15th-century Church of the Holy Trinity, which the local people claim with pride to be the finest, lightest and most attractive church in Suffolk, which is saying a great deal. Nearly 100 high windows contribute to the airy feel of the building which is built in the Perpendicular Gothic style and dates from 1450.

Near the green lies Long Melford Hall, an early-Elizabethan manor house with Georgian additions which include pepperpot towers. This was a hunting lodge for the abbots of Bury St Edmunds, and after it passed into the hands of William Cordell, a counsellor of Henry VIII, it became one of the few places in England where Elizabeth I definitely spent the night during the Royal Progress through Suffolk in 1588. Since 1786 it has been the home of the Parker family, who open the house and gardens to visitors throughout the summer.

LONG MELFORD lies on the A134, 10 miles south of Bury St Edmunds.

MALDON
Essex

Little Maldon, once a Roman seaport, lies at the head of the Blackwater estuary, a pretty place, full of old buildings and popular today with yachtsmen. The town is commemorated from the 10th century in an Anglo Saxon epic poem, *The Battle of Maldon*, and was the first town in Essex to gain a charter from the Norman kings in 1171. It continued to serve as a busy seaport throughout the Middle Ages and this period has left the town with such places as the Moot Hall, built in 1450, and still used by the Council, the old leper hospital, the Hospital of St Giles, the Blue Boar Inn which is early 16th century, and the Church of All Saints. Maldon is a lovely spot, the ideal place to spend a weekend away from London, loud with the cry of gulls, where the air is full of the salty tang of the sea.

MALDON lies 10 miles east of Colchester on the A414.

NEWMARKET
Cambridgeshire

It could probably be said, with a considerable degree of accuracy, that Newmarket is the horseracing capital of the world. Certainly it has antiquity on its side, for it was the Scots nobles of James I who introduced horseracing to England in about 1605, and they found the springy turf of Newmarket ideal for these contests.

Charles II adored horseracing, and made Newmarket his second home. The Rowley Mile course at Newmarket is named after the King's

The Royal Arcade in Norwich is an excellent shopping centre

hack, Old Rowley, and indeed was the King's nick-name. The town was well established as a training and racing centre by the end of the 17th century and has remained that way ever since. The Jockey Club maintains its headquarters in the High Street, racing stables are everywhere, and the normal sound early in the morning is the clip-clop of hooves as the strings move out to the gallops. As a result of all this, there are good pubs, fine restaurants, and a generally pleasant country air about the place.

NEWMARKET lies 15 miles west of Bury St Edmunds on the A11.

NORWICH
Norfolk

Norwich — a very fine city. That's what the tourist board claims and few would disagree with that firm statement. It is an old place with Saxon roots, a Norman castle, a magnificent cathedral, and a pub for every other day of the year.

Like many of England's older cities, it is clogged by the motor car, and really must be seen on foot, so be sure to explore the Haymarket, stepping into the glorious church of St Peter Mancroft, then on to see Charing Cross, a medieval merchant's house, and for an historic over-view of the city, stop for a tour of the Bridewell Museum. The city is full of good architecture from almost every period, but the houses which line cobble stoned Elm Hill are certainly among the most picturesque.

Norwich Castle is a must, and especially so now it contains the County Museum. It was built in the 12th century, and then became the city jail, which kept it intact until 1887. It has recently been fully restored and is now a very splendid castle indeed, with a museum and art collection.

Norwich Cathedral is one of the finest in England, the pride of this fine city. It was begun in 1096, and with many later additions and alterations is a magnificent building with a notable interior.

The vault of the nave is studded with ceiling bosses, and the Bishop's Throne is said to pre-date the Conquest. Outside the cathedral two sights to see are the grave of Nurse Edith Cavell, shot by the Germans in 1915, and the Erpingham Gate which contains a carving of old Sir Thomas Erpingham who fought at Agincourt and features in Shakespeare's *Henry V*. Explore Norwich thoroughly — it really is a fine city.

NORWICH lies 19 miles east of Great Yarmouth on the A47.

OTLEY HALL
Suffolk

If the Georgian style is the most attractive form of urban dwelling, the medieval manor house fills the same proud position out in the countryside.

Otley Hall, nine miles north of Ipswich, is a perfect example of this attractive genre. It was built in the latter part of the 15th century and today remains in superb condition and is a splendid example of English domestic architecture. It has a moat, tall pointed chimneys, dull-red herringbone brickwork, oak beams, wall paintings, pargetting and panelling — in short the lot. It was built by the Gosnold family, who lived in it until the mid-18th century, during which time one member of the family, Bartholomew Gosnold, a sea-captain, crossed the Atlantic to discover Cape Cod and found Martha's Vineyard. The manor house is open to visitors throughout the summer and is well worth a visit.

OTLEY lies 9 miles north of Ipswich, off the B1079.

The Norfolk Broads

For all its concentration on agriculture, East Anglia is a great holiday region, containing in the flat country between Norwich and the North Sea fishing port of Great Yarmouth, an attraction which is unique to England; the Norfolk Broads.

This area of fen and marsh, lined by thick reed beds, dates back to the Middle Ages, when the local commoners exercised their rights to cut turfs from any 'turbury' or turfland. Because of extensive turf cutting, broad channels were created and then flooded by the sea and the overspill of such local tidal rivers as the Yare, Waveney, Buse and Ant. The main flooding occurred at the end of the 13th century and by the 16th century the present Broads had already come into existence, a haunt of heron, and the booming bittern, full of duck, a paradise for wildlife.

Today the Norfolk Broads are a water-lover's haven, full of boats in summer, with most of the action centred around the inland port of Wroxham, where the rows and rows of cabin cruisers can be seen each summer, tied up, waiting for their crews or chugging out into the stream, but not just modern cruisers; there are sailing dinghies and sailboards and small yachts, even from time to time a stately Norfolk wherry, beating against the tide with her mass of high tan-coloured sails.

Today, although much eroded and under the ever-present threats of silting up and man-made pollution, the Broads remain a unique part of the English landscape, and are well worth a visit at any time of year.

As one of the largest remaining wetlands in England, they are full of rare plants and provide a habitat for many rare resident and migrant birds, so that those who love natural history will find the Norfolk Broads an experience as well as a playground.

A yacht cuts its way across Hornsey Mere, behind the reed beds, under the high blue sky

THE OTTER TRUST
Earsham, Suffolk

Over the last few decades, that endearing creature, the otter, has become one of the world's most threatened species. Here, in a conservation centre near Bungay in Suffolk, the otter is being conserved, bred and re-introduced to the wild. The centre is devoted to the conservation of otters from all parts of the world, but particular attention is paid to British otters. The otters are kept in large natural enclosures, each with a stream for them to play in. Otters apart, there are three large lakes with great numbers of waterfowl, and a wood full of muntjac deer. An attractive and useful place which will attract all visitors passing through Suffolk.

EARSHAM lies 2 miles west of Bungay off the A143.

PETERBOROUGH
Cambridgeshire

Peterborough is a neglected cathedral city, which is a great pity, for it has a lot to offer the dedicated traveller with a sense of history. It is now a noted centre for heavy engineering with an excellent shopping centre, and has been a prosperous place since Norman times. In spite of much modern development on the outskirts, those who penetrate to the city centre will find the old 17th-century market place quite delightful, while Priestgate contains many fine old houses. The glory of the town though, is Peterborough Cathedral, which may not rank with the other cathedrals of Britain, but is still the most impressive Romanesque building in Britain, and will please any lover of the Romanesque style.

Architecture apart, it is full of history. The long-suffering Catherine of Aragon, wife of Henry VIII, is buried here, as was Mary, Queen of Scots after her execution at Fotheringhay, before her body was removed to Westminster Abbey. The bulk of the cathedral dates from the 12th century, but the West Front is 13th century, as is the nave and much of the decoration comes from the High Medieval period.

PETERBOROUGH lies 35 miles north of Cambridge, off the A1.

Pargetting (patterns pressed into plaster) is a feature on a number of buildings in Saffron Walden

SAFFRON WALDEN
Essex

To those who love that classic domestic architecture which dates back to Tudor and medieval times, the town of Saffron Walden is a necessary visit, for the town contains an almost unique collection of beautiful buildings. Besides, the name itself is attractive, and dates back to the 14th century when the saffron crocus grew in quantity hereabouts.

On a small rise above the common, stands the ruin of a castle built by one of the Conqueror's comrades, Geoffrey de Mandeville, in about the year 1100. That, inevitably, was destroyed by Cromwell, who quartered himself at the Old Sun Inn in the town centre during the siege. The church of St Mary is the largest in Essex and is built in that soaring Perpendicular Gothic, a triumph of verticality, and the vault and interior carvings are equally magnificent.

After visiting the town, go out a mile to the west, to Audley End House. This was once a Benedictine monastery, dissolved and destroyed at the Dissolution, and the ruins sold to Sir Thomas Audley who built the first house here before the estate passed to Lord Howard de Walden. This house in turn was largely pulled down at the start of the 18th century when Sir John Vanburgh, the architect of Blenheim Palace was in charge of the work. Alterations continued for most of the century and the result is the Audley End House of today, a mixture of styles reflecting the work of various masters; John Vanburgh, Robert Adam, and gardens of course by Capability Brown. Much smaller today than in its Jacobean heyday, Audley End House remains magnificent, full of paintings and fine furniture.

SAFFRON WALDEN lies 1 mile east of the M11, north of Bishop's Stortford.

SOUTHWOLD
Suffolk

An attractive, elegant Victorian town set on a cliff top over a shingle beach, Southwold has always been a busy fishing port. For the last 100 years or so it has also been a popular seaside resort, but still maintains a life of its own outside that supplied by the summer visitors.

There are fishermen's cottages, fine old streets, good pubs, notably the Sole Bay Inn, and a church dedicated to St Edmund, that popular East Anglian saint, with a good example of Perpendicular architecture. One feature of the church is a manikin, Southwold Jack, which dates back to the 15th century and strikes the bell of the church clock. The town museum, which contains relics of the Anglo-Dutch naval battle of Sole Bay of 1672, is an interesting place, but the chief attraction of Southwold lies in strolling about the old streets and taking the air along the beach.

SOUTHWOLD lies 16 miles south of Lowestoft, off the A12.

Southwold's elegant houses face the green

SUDBURY
Suffolk

Although Suffolk is best known among art lovers for the 'Constable Country', it produced at least one other famous English painter, Thomas Gainsborough, who was born in Sudbury in the elegant house in what is now Gainsborough Street, which contains the Gainsborough Museum. The contents include work by this artist and also by Constable, much 15th-century furniture, and Gainsborough memorabilia.

Sudbury itself is one of those market towns which look so right in the green parts of the English countryside. It is an ancient town, full of fine buildings and contains one good example of a medieval church, the 14th-century St Gregory's, where the church treasury contains a curious relic of another of Sudbury's sons, Simon Sudbury, a local cleric who rose to be Archbishop of Canterbury under Richard II. During the Peasants' Revolt in 1381 he was seized by the rioters and beheaded on Tower Hill, but his skull somehow found its way back to Sudbury and can be seen in St Gregory's Church.

SUDBURY lies 23 miles west of Ipswich on the A1071.

SWAFFAM
Norfolk

Swaffam, with a population of about 4000, is rather more a large village than a town, but it is still a busy, thriving and beautiful town of the 18th century, and as such, well worth a visit. The market square is actually a triangle, and lined with fine buildings looking out at an unusual Palladian style market cross under a cupola and set on columns. The parish church pre-dates most of the town and was built in the 15th century. It is a splendid Perpendicular building with a notable hammerbeam roof. Of the attractive Georgian style which dominates the rest of the town, two of the finest examples are the old School House and the Assembly Rooms, both worth inspection.

SWAFFAM lies on the A47 road, 28 miles west of Norwich.

THAXSTEAD
Essex

Although the country's steel and cutlery trade is commonly supposed to reside in Sheffield, Thaxtead craftsmen have been producing fine English cutlery since the Middle Ages. In 1380 no less than 89 cutlers or blacksmiths were working in the town, and they built the Guildhall in Town Street to house the Cutlers Guild just ten years later. As a result of this industry, which thrived until the 16th century, Thaxstead is a rather splendid town, topped off by the soaring spire of the town church, St John's, an excellent medieval church, decorated with gargoyles and glass. Between here and the Guildhall the streets are lined with 15th-century houses.

THAXTEAD lies 9 miles south-east of Saffron Walden.

WALSINGHAM
Norfolk

Walsingham is a haunting spot, for centuries the site and objective of the most famous and popular of all the English pilgrimages, to the shrine of Our Lady of Walsingham. Henry VIII came here before he ordered its destruction; Sir Walter Raleigh lamented the glory that was gone. Not much remains to see today, although a fresh shrine has been established by the Catholics. The Walsingham Abbey grounds contain the former site of Our Lady of Walsingham, with the remains of the Augustinian priory and the old gatehouse. The present shrine of Our Lady is in Holt Road nearby, but the medieval pilgrimage can be recalled at the small 14th-century Slipper Chapel at Little Walsingham, two miles to the south, where pilgrims would wash their feet before walking barefoot to Walsingham. You can still do that today, and many do.

Pilgrims still visit the shrine of Our Lady of Walsingham

WALSINGHAM lies on the B1105 road, 4 miles south of Wells-next-the-Sea.

WELLS-NEXT-THE-SEA
Norfolk

Wells is a delightful, old fashioned Norfolk fishing port, famous for whelks and silvery sprats. It is quite large, full of flint-stone houses set around the Green, with a main street running down to the quay. A wide beach runs off beside the harbour wall, backed by pine trees. Many of the houses are Georgian and a surprising number seem to be pubs. It is still a port for small coasting craft, and is at one end of a little narrow-gauge railway which runs inland to Walsingham.

WELLS-NEXT-THE-SEA lies on the A149, 4 miles east of Burnham Market.

WROXHAM
Norfolk

What Wroxham is, hardly matters. It is a centre for that magical place, the Norfolk Broads. That fact cannot escape even the passing eye in Wroxham, for the little town is a miniature Venice, completely surrounded by water, and the surface groans with pleasure craft, cruisers, yachts and motor boats. This town is the place where holidaymakers equip their craft for a holiday on the Broads, and it is a hive of activity in summer, a busy, noisy spot, full of colour. Wroxham Barns, built in the 18th century, contain a unique display of rural crafts, a museum and such working exhibits as a wood carver and a potter.

WROXHAM lies 9 miles from Norwich on the A1151.

DERBYSHIRE
NOTTINGHAMSHIRE
LINCOLNSHIRE
NORTHAMPTONSHIRE
Mam Tor
Eyam
Chatsworth House
Hartington
Tissington
The Tramway Museum Crich
Thoresby Hall
Newark Castle
Southwell
Nottingham
Lincoln
Louth
Boston Stump
Grimsthorpe Castle
Bourne
Spalding
Crowland Abbey
Melton Mowbray
Ashby-De-La-Zouch
Bosworth Field
Oakham
Burghley House
Rockingham Castle
Deene Park
Oundle
Broughton House
Naseby
Higham Ferrers
Althop House
Northampton
The Waterways Museum Stoke Bruerne
A 631
A 15
A 46
A 16
A 1
A 57
A 158
A 17
A 52
A 47
M 1
M 69
0 10 20 30 km
0 10 20 miles

The English Shires

This peaceful scene at Foxton Locks, Leicestershire, typifies the scenery to be found within the English Shires, much of it unchanged for centuries

WITH THE CERTAIN exception of the Derbyshire Peak District, the English Shires have somehow slipped through the tourist net. Despite the fact that they have a great deal to offer the traveller, in the shape of fine towns and a beautiful countryside well laced with history and splendid country houses, this, the very heart and heartland of England, is all too often overlooked.

What links these counties together is the life on the land, which to a very great extent has gone on unchanged for centuries. These are the hunting shires of England, which come alive each autumn to the clip-clop of hooves on cobble-stones, the high cries of the hounds and the long drawn-out wail of the horn — and whatever you may think of hunting the fox, few can deny that the sight of the hunt in full cry, thundering across the fields in red coats and a spattering of mud, is a thrilling and memorable spectacle.

Within all this farming country, tucked into the patchwork of field and copse and forest, lies the Derbyshire Peak District, one of the glories of the English countryside. Visitors here will catch a touch of the old pagan past of England in such ceremonies as the Tissington Well Dressing, or recall the tragedy of the Great Plague in nearby Eyam, where the population elected to stay in their village and die when the plague came, rather than scatter and carry the plague about the country; a quiet, very English sort of courage.

ALTHORP HOUSE
Northamptonshire

This splendid house is owned by the Spencers — the Princess of Wales' family. It was built by Sir John Spencer in 1508, rebuilt by Henry Holland in 1790, and redecorated from top to bottom as recently as 1982. It is still the home of the present Earl and Countess Spencer but open to visitors throughout the summer. Apart from the curiosity of seeing inside the home of the country's future Queen, there is a magnificent art collection, with many works by Reynolds and Gainsborough, excellent French and English porcelain, and some rare items of furniture. The park is very fine and the nearby village most attractive.

ALTHORP lies near Harlestone, 6 miles north-west of Northampton on the A428.

ASHBY-DE-LA-ZOUCH
Leicestershire

Most people remember Ashby-de-la-Zouch from Sir Walter Scott's novel *Ivanhoe*, where it appears as the site of the famous tournament where the mysterious knight first makes his appearance and overthrows the Templars. Ashby today is a quiet, rather pleasant little town with many Georgian buildings and little evidence of the Middle Ages, except the ruins of Ashby-de-la-Zouch Castle, which dates from the early days of the 12th century and was later extended and strengthened by Lord Hastings before he was executed for treason by Richard III. During the Civil War the garrison held this castle for the king for over a year, but after they surrendered, with all the honours of war, the Parliamentary army 'slighted' the walls, and reduced the castle to its present ruined condition.

The ruins of the Norman Castle are all that remain of the fortifications of Ashby-de-la-Zouch

ASHBY-DE-LA-ZOUCH lies 20 miles north-west of Leicester on the A50.

BOSTON STUMP
Lincolnshire

Standing out above the roof tops of this old port and market town, the tall tower — or stump — of St Botolph's Church is a landmark for miles around. At 272 ft high the top offers marvellous views to those strong enough to plod up the 365 steps, one for every day of the year, which lead out onto the roof. The Stump is an airy church, well furnished and full of interest, as is the little town which surrounds it. Boston, on the edge of the Wash, gained its charter from King John in 1204, and was a major port for the cloth trade throughout the Middle Ages before the harbour began to silt up and the balance of trade switched to the West Coast ports. It was from here in 1612 that a group of Puritans sailed for America and founded the great city of Boston in Massachusetts. Around the town today visitors can inspect many

The tall tower of St Botolph's Church, better known as Boston Stump

16th and 17th-century houses, the 15th-century Guildhall (now a museum) and the local theatre which occupies the half timbered Shodfriars Hall in South Street.

BOSTON lies 25 miles north of Spalding on the A16 and 34 miles north west of King's Lynn.

BOSWORTH FIELD
Leicestershire

On the banks of the Ashby Canal, close to the town of Market Bosworth, lies Bosworth Field, where in 1485 Richard III, the last Plantagenet King of England, was killed, '......fighting valiantly in the press of his enemies'. His crown was retrieved from a thorn bush and placed on the head of Henry Tudor, and a new dynasty came to the throne of England. The battlefield is marked today by a Battlefield Visitor Centre with exhibitions and models, a film theatre where the Olivier version of Shakespeare's *Richard III* is often shown, and the main features of the battle can be followed on the waymarked Battlefield Trail. A bookshop and cafeteria complete the amenities, and the whole area is well planned and well worth a visit. Richard III has been greatly maligned, not least by Shakespeare, so it is worth remembering that on the evening of the battle a chronicler of York wrote in the city journal, 'This day was our good King Richard slain, to the great sadness of this city.

THE BATTLEFIELD lies near Sutton Cheney, 2 miles south of Market Bosworth, off the A447.

BOURNE
Lincolnshire

Bourne is a place of legend in the Fen Country, the birthplace of that Saxon warlord, Hereward the Wake, who retreated into the Fens after Hastings and fought off the Normans for years — or so it is said. In fact, the evidence for Hereward's existence is slight, and the evidence for his connection with the De Wake family who built the Norman castle here and founded the Abbey in 1138 is virtually non-existent. Inevitably, Bourne Abbey was destroyed by Henry VIII, the Great Vandal, but the Abbey Church survived as the parish church, and still stands. Note Red Hall in South Street, where lived the Digby family, executed for their share in the Gunpowder Plot of 1605. The Red Hall eventually became the home of the local station master, and is now a museum with some fine Elizabethan furniture.

BOURNE lies 12 miles west of Spalding on the A151.

BROUGHTON HOUSE
Kettering, Northants

This splendid Tudor mansion began its existence as a monastery. The central parts date back to the mid-1450s and were gradually enwrapped in a Tudor building, spread over seven courtyards and completed in 1695 with an addition in the style of Louis XIV's palace at Versailles, which looks both unusual and attractive. The house now contains some priceless works of art, including paintings by the Spanish masters, El Greco, (who was actually Greek), Murillo and Caracci. There is a notable private collection of arms and armour, a coach house, excellent 17th and 18th-century furniture in both the English and French styles, fine ceilings and, to top it all, the most beautiful gardens. 'House' is hardly the word for it — Broughton is really a palace.

BROUGHTON lies 3 miles north of Kettering off the A43.

BURGHLEY HOUSE
Stamford, Lincolnshire

Burghley was built by Elizabeth I's great counsellor, William Cecil, Lord Burghley, and completed in 1598, only three years before the Queen died. The house was badly damaged by Cromwell's forces during the Civil War, but a great deal remains of the Elizabethan building, including the Great Hall. The beautifully proportioned State Apartments contain fine furniture, paintings, silver and tapestries, while the ceiling of the Great Hall is itself a masterpiece. Verrio painted the Heaven Room in 1694 and the gardens are the creation of Capability Brown, who laid them out between 1760 and 1780. The park is open throughout the summer and in early September is the site of the famous Burghley Horse Trials.

BURGHLEY HOUSE lies on the B1443, 1 mile south-east of Stamford.

CHATSWORTH HOUSE & GARDENS
Derbyshire

Chatsworth House, gardens, and collections are just simply magnificent. Chatsworth is more like a palace than a country house and has been the home of the Dukes of Devonshire since the 17th century. The present house stands on the site of a Tudor mansion and is built entirely from local materials, grit-stone, marble and black-stone, all from nearby quarries. The house is full of treasures, furniture, paintings by Poussin, Rembrandt and Reynolds, and some exquisite porcelain, but the park and gardens really draw the crowds. The famous arboretum contains trees from all parts of the world,

 Chatsworth House in autumn shades

the Grand Cascade is the largest of many waterfalls and the Emperor Fountain, built to commemorate the visit of Tzar Nicolas I can throw a jet of water nearly 300 ft into the air. The gardeners at Chatsworth are clearly devoted to their work, for with its shrubs, roses, trees and floral beds, the garden is a picture.

CHATSWORTH lies 3 miles from Bakewell on the A619.

CROWLAND ABBEY
Lincolnshire

Crowland Abbey, or what remains of it, lies in the Fen Country of Lincolnshire, just on the border with Cambridgeshire. The original abbey was founded in the 8th century, and destroyed by the Vikings in AD 870. The then abbot's skull, still marked by a sword blow, is displayed in a glass case on the north wall of the Abbey Church. Constant raids and the eventual passing of Crowland into the Danelaw, virtually extinguished the Celtic abbey and the present one dates from the 12th century, after which Crowland grew in prosperity up to just before the Dissolution, when one of the abbots was Henry VII's chancellor, Bishop Morton of 'Morton's Fork' fame.

Henry VIII's commissioners destroyed most of the abbey in 1538 and what remained was damaged still further during the Civil War. What can be seen today is still attractive, the north aisle of the church and the West Front set off by the arch of the Norman abbey.

CROWLAND lies 10 miles south of Spalding on the A1073.

DEENE PARK
Northamptonshire

Deene is a tiny village with a population hovering around the hundred mark, set in a leafy valley in the limestone northern part of the county. The great attraction hereabouts is Deene Park, a classic house in that beautiful light-coloured Welden stone, owned for several centuries by the Brundells, one of whose ancestors led the Charge of the Light Brigade at Balaclava during the Crimean War. Sir Robert Brudenell bought Deene Park in 1514 and it is a fine building with Tudor roots, a deer park with a large lake and extensive gardens, full of flowerbeds and shrubs, all in all a delightful spot.

DEENE lies 8 miles north-east of Corby on the A43.

EYAM
Derbyshire

Eyam is a beautiful Derbyshire village with a sad but historic past. During the time when the Great Plague was sweeping London in 1665, a local tailor received a contaminated consignment of cloth from the city. The bubonic plague swept through the village and although flight into the surrounding countryside would have saved lives, it would also have risked spreading the disease throughout the country. Led by their vicar, the villagers of Eyam decided to stay put. Within a year 259 of the 350 villagers were dead and their graves can be seen around the village in little clumps. The six 17th-century plague cottages where the victims were housed still stand, and the tragedy is commemorated with a procession on the last Sunday in August. Other sights nearby are Monpessons Well, named after that gallant vicar, and on a nearby hill a Celtic preaching cross. Today Eyam is a pretty spot and a pleasant place for the visitor to wander about.

EYAM lies off the A623, 10 miles west of Chesterfield.

The graves of the Hancock family, Eyam plague victims

Well-dressing in Derbyshire

Though well dressing is not entirely restricted to Derbyshire, it is seen at its best in a handful of villages which lie deep in the folding valleys and hills of the Peak District, at Tideswell, Tissington, Barlow, Wirksworth and Youlgrave.

Although the origin of this custom is almost certainly pagan, the ceremony has been Christian since the Middle Ages, and is said to be a direct link with the Black Death, when a supply of unpolluted water was essential to the survival of small communities.

Those who have never seen the beautiful pictures and patterns which the villagers form by each well, pressing flower petals into flat trays of wet mud, will be amazed at the detail, and the artistry, for these 'portraits in petals' illustrating religious themes or scenes from the Bible are works of art indeed.

The well dressing takes place on Ascension Day each year, and the villagers compete to make their well, or wells, the most beautiful in the district. Tissington has five wells, and each is covered with a large wooden frame, filled with wet clay and then carefully decorated.

At 11 o'clock on the morning of Ascension (the Thursday before White Sunday — now the Spring Bank Holiday weekend) a procession led by the clergy, will visit and bless each well.

Derbyshire well dressing, seen here at Tideswell

GRIMSTHORPE CASTLE
Lincolnshire

Grimsthorpe is a dramatic place and it has seen dramatic events, not least in 1541 when Henry VIII chose a tournament at Grimsthorpe as the place to accuse his wife Catherine Howard of adultery. Poor Catherine lost her head, and their host, the Duke of Suffolk, was covered with understandable confusion. That scene was preceded by some rapid rebuilding to house the court, and only one tower dates from the original 13th-century building.

In the early 18th century, Vanburgh, the architect of Blenheim, was asked to undertake the conversion of Grimsthorpe into a palace, but although he died long before the work was completed, Grimsthorpe is certainly the greatest country house in Lincolnshire. The Great Hall is magnificent, and the castle contains such interesting relics as the suit Charles I wore for his portrait by Van Dyke, and some fine paintings, (including Van Dyke's) and much beautiful furniture.

GRIMSTHORPE lies west of Bourne on the A151.

HARTINGTON
Derbyshire

Hartington is the classic little village of the Peak District, a good touring centre for the entire area and an attractive place in its own right. Places to visit include the Charles Cotton Inn, named for Izaak Walton's famous fishing companion, and the local Stilton cheese factory, which has that delicious, tangy cheese for sale in attractive little pots. The Market Place is enhanced by a wide green and a duck pond, and is surrounded by pretty houses and cottages. Close by, and a favourite excursion for visitors to Hartington, lies the beautiful Dovedale, marking the boundary between the Derbyshire and the Staffordshire Peaks.

There are good pubs and cafés, a youth hostel and a small hotel, all the necessities to make a visit to Hartington worthwhile.

HARTINGTON lies off the A515, 10 miles north of Ashbourne.

HIGHAM FERRARS
Northamptonshire

The Market Square, composed of beautiful stone houses grouped around the early 19th-century Town Hall, provides the central gem of Higham Ferrars. Today it is a pleasant, busy market town, but it has deep historic roots. Close to St Mary's parish church stands the medieval Bede House and Chantry. In the Lady Chapel of St Mary's the East Window commemorates Henry Chichele, who was born in Higham Ferrars and rose to become Archbishop of Canterbury in 1413.

HIGHAM FERRARS lies 10 miles east of Northampton, on the A45.

LINCOLN
Lincolnshire

Few cities in England can compare with Lincoln, a place where the setting and the architecture have conspired to create a town of rare beauty and a distinctly different atmosphere. Lincoln reeks of history; the Romans came this way and travellers still enter the city under the Roman Newport arch. Inside, the streets are lined with old houses, some dating back to Norman times, leading on to the half-timbered Tudor buildings and shops that stand on the medieval bridge over the river. Above all this, and visible for miles across the flat surrounding countryside, stands Lin-

High on a hill above the city stands Lincoln Cathedral

coln Cathedral and Castle set on the hilltop which rises above the city. Lincoln Cathedral is one of the largest in England, floodlit at night like a great stone ship. The first cathedral here was built in 1092, but this was destroyed by an earthquake in 1186 and the present building was not finally completed until 1280. It contains all the glories of the English Gothic, and the interior is noted for the carved choir stalls, the 13th-century glass, the Angel Choir and the large collection of medieval brasses.

The castle, a fearsome pile, was built by William the Conqueror in 1076, although the gatehouse is much later, dating from the 14th century, while the keep, or Lucy Tower, was built about 1130. The castle was a royal prison, and the old cells and even iron fetters can still be seen, notably in the prison chapel.

LINCOLN lies 16 miles north-east of Newark on the A46.

THE LINCOLNSHIRE BULB FIELDS

Spalding

The flat fen country around Spalding and Holbeach in the south of Lincolnshire is the centre for one of England's most prosperous and attractive rural industries — the growing of tulips and daffodils. Springfields at Spalding is the centre for this activity, and in the springtime the fields around both towns are a multi-coloured carpet as the flowers come into bloom. In May, Spalding holds a Flower Festival when flower-decked floats parade through the streets to mark the end of the bulb season and the start of the summer, when roses and bedding plants make their colourful contribution to the local landscape.

SPALDING lies 20 miles south of Boston on the A16.

LOUTH

Lincolnshire

Louth is the capital of the Lincolnshire Wolds, a broad strip of country lying across the north-east corner of the country, designated an Area of Outstanding Natural Beauty, and crossed by one of Britain's most popular footpaths, the Viking Way. Louth stands on the eastern side of the Wolds, and is the main market centre for the district, being once second only to Lincoln. Today, Louth is a dreamy little place, mostly Georgian but topped by the splendid Church of St James, which has a 140 ft spire and is definitely one of the most beautiful churches in England, built between 1450 and 1515. The great Cistercian Abbey, which once provided the town with wool for the

The medieval home of Aaron the Jew in the historic heart of Lincoln

thriving local cloth trade — that Lincoln Green so popular with Robin Hood — was destroyed at the Dissolution when many of the monks were hanged in the town square for refusing to obey the Royal Commissioners. As a touring centre for the Wolds, Louth is quite ideal and not far away lies the long, curving attraction of the Lincolnshire Coast.

LOUTH lies 27 miles east of Lincoln on the A157.

MAM TOR
Derbyshire

Derbyshire is a county best known for its superb countryside, which must rank among the most attractive in England, with jumbled dales, steep hillsides lined with drystone walls, rushing streams, grit-stone rock faces, and the endless tugging of the wind. In a county which abounds in great views, one stands out above the rest. Mam Tor is a lofty, windy spot on the B6061 road, crowned at the very top by the relic of an Iron Age hill fort. Above here, the hang-gliders swoop and hover, and lesser mortals can look out across Castleton far below in the valley to Edale, and the southern end of the Pennines. Castleton lies in the so-called White Peak, a limestone region riddled with vast underground caves.

MAM TOR lies 1 mile west of Castleton on the B6061.

MELTON MOWBRAY
Leicestershire

Melton Mowbray is yet another of those old, traditional market towns, a centre for the local countryside and country people. It is particularly famous for the quality of its local pork pies which are said to contain 'everything but the squeal' of the pig, and for the manufacture of Stilton cheese. Both products can be purchased in the town and their history is recorded in the Carnegie Museum. Melton Mowbray is also one of the great hunting towns, the home of the famous Quorn Hunt, and the clip-clop of hooves is almost as common a sound here as the roar of traffic. The town has a fine parish church, with every shade of Gothic, and a very busy open air market. In the surrounding countryside lies the beautiful Vale of Belvoir, another famous hunting region, well worth exploring, but do not fail to spend a day in Melton Mowbray, a town well equipped with old ruins, fine shops and interesting sights.

MELTON MOWBRAY lies 15 miles north-east of Leicester on the A607.

NASEBY
Northamptonshire

The Battle of Naseby in 1645 effectively settled the outcome of the English Civil War, for here Cromwell's New Model Army shattered King Charles' Cavaliers, and destroyed the King's power for ever. The Battle and Farm Museum contains a miniature layout of the battle as well as farm tools and implements, while the battlefield itself is marked with an obelisk. This is best visited after hearing the commentary on the battle in the museum.

THE BATTLEFIELD lies off the B4036, 20 miles north-east of Daventry.

NEWARK CASTLE
Nottinghamshire

Newark has been voted one of the most historic towns in the country, but it is still very far from a tourist trap, in spite of having a fine castle, a popular racecourse and a great many old buildings.

St Mary Magdalene Church, in the market place, has a tall spire, many curious gargoyles, and a notable brass of 1349. There are a great many fine old pubs, notably the Saracen's Head and the Clinton Arms, both frequented in their time by Sir Walter Scott. The houses in Kirkgate are even older, being Tudor, while Newark Castle is older still.

The castle was built to secure the Great North Road in the 12th century, and severely damaged during the Civil War when it was besieged by Parliament. The walls and towers

Despite damage during the Civil War, Newark Castle is spectacular

are still intact though, and the 12th-century chapel can be visited. An attractive town, quite unspoiled, Newark is one of the places which must find a place in the itinerary of a visit to the Midlands.

NEWARK lies 8 miles east of Southwell.

NORTHAMPTON
Northamptonshire

No visitor to Northamptonshire can ignore the county town, once a centre for the shoe trade and very historic under the usual modern trappings. King John gave Northampton Castle to the Barons as a pledge for keeping Magna Carta, though Charles II, who disliked the inhabitants' Parliamentary attitudes, had it and the town walls pulled down at the end of the 17th century. During the Wars of the Roses, a battle took place here in 1460, when the Lancastrians were defeated and Henry VI taken prisoner. At nearby Hardingstone travellers can inspect one of the few remaining Eleanor crosses erected across England by Edward I to mark the places where the body of his beloved wife, Eleanor of Castile, rested on its way from Nottinghamshire to London. There is another, even finer, near Geddington. Back in the town, St Peter's Church is Norman, and All Saints Church early medieval. The town museum includes an exhibition of leathercraft and the shoe trade, and the shopping is excellent.

NORTHAMPTON lies 4 miles east of the M1, off Junction 15 by the A43.

NOTTINGHAM
Nottinghamshire

The old city of Nottingham is said to have the prettiest girls in England and it may well be true. That apart, it is a most attractive town. Any visit might begin with a drink at the famous Trip to Jerusalem Inn by the castle, from which Crusaders set out for the Holy Land. The castle itself was destroyed in the Civil War, but 'The Trip' is still in business and is said to be the oldest inn in England, dating from 1189.

Horseshoes paid as toll fees in the Banqueting Hall of Oakham Castle

Close by stands the statue of Nottingham's most famous son, Robin Hood, who warred for justice with the notorious Sheriff, while for something a little more accurate than those legends of the past, visitors might take in the Museum of Costume at Castlegate which displays among other items a 200 year old collection of underwear.

The old houses of the lace-makers, off the Lace Market, are very attractive, as are the shops surrounding the great central market area where Nottingham's famous Goose Fair is held every October.

Today Nottingham is famous for bicycles, and for good pubs, and is a most attractive city for visitors and residents alike.

NOTTINGHAM lies 50 miles north-east of Birmingham.

OAKHAM
Leicestershire

Oakham was once the county town of the now-vanished but well-remembered little county of Rutland, which Leicestershire absorbed in the 1970s. Oakham is a very fine town, with a famous public school, Uppingham, an excellent county museum and a number of historic and attractive inns. One somewhat different local attraction at Oakham is the Rutland Farm Park, where rare breeds of sheep and cattle are collected, conserved and bred, and can be seen cropping the grass under the trees. Oakham Castle is Norman and visitors should note the numerous horse-shoes which adorn the walls of the Banqueting Hall. These were paid over as a toll fee by any baron who

A detail from the exquisite foliage carving on a window in the Chapter House at Southwell Minster

passed a night in the castle. Other sights include the old market square, Oakham School, and not far away, the vast sweep of Rutland Water, a reservoir as big as Lake Windermere, ideal for sailing, fishing or birdwatching.

OAKHAM lies 11 miles west of Stanford on the A606.

OUNDLE
Northamptonshire

Oundle is a very fine little town, a beautiful spot, and rightly designated as a Conservation Area. The stone buildings which make it so attractive include those of the public school.

The White Lion Inn dates from 1641 just before the Civil War, while much of the Talbot Inn was built with stones filched from the ruins of nearby Fotheringhay Castle, where Mary, Queen of Scots was executed. Needless to say, the Queen's ghost is said to haunt the inn. The parish church is early medieval with fine glass and brasses, and the almshouses were endowed in 1611.

OUNDLE lies 8 miles east of Corby on the A427.

ROCKINGHAM CASTLE
Northamptonshire

If Rockingham looks familiar to the visitor that may be because it swept to worldwide fame as Arnscote Castle in the TV series *By the Sword Divided*. Descriptions may therefore be superfluous, for it is clearly a beautiful place. William the Conqueror built the first castle here, and the present building has been the home of the Watson family since 1530. Their house is basically Elizabethan, but the outer walls are Norman and encircle such medieval relics as a tilt yard, a 16th-century yew hedge and fine rose gardens. The interior has a good collection of Rockingham china, excellent paintings and a notable connection with Charles Dickens. The castle is open in summer and should certainly be included during any visit to this part of the country.

ROCKINGHAM lies 10 miles north-east of Market Harborough on the A6003.

SOUTHWELL
Nottinghamshire

Southwell is a splendid place, overlooked and beautified by the soaring magnificence of Southwell Minster. This was built by the Normans and only elevated to cathedral status as recently as 1884. The brass lectern in the Minster is all that remains of the contents of the nearby abbey at Newstead, largely destroyed at the Dissolution, but even James I remarked that the Minster, which survived this unhappy time, was '...the finest kirk in England'.

Byron entertained his ladies at the Saracen's Head Inn, and it was here that Charles I surrendered himself to the Scots in 1646. Mostly though, it is the great Minster which attracts the visitors today and rightly so, for it is most decidedly a splendid, evocative building.

SOUTHWELL lies 8 miles west of Newark on the A612.

THORSEBY HALL
Worksop, Nottinghamshire

Thorseby Hall isn't as old as it looks. It appears to be a Tudor building but actually dates from 1864, and is the result of a Neo-Tudor enthusiasm which swept through England. It has a hammer-beam roof, an intriguing clock collection, various state apartments, and over 200 rooms; no one is quite sure exactly how many, but that is one estimate. The park is equally vast, covering over 12,000 acres. In the 1800s the house was the home of Lady Mary Montagu, a writer and traveller who pioneered the concept of innoculation against smallpox.

THORSEBY lies 7 miles south-east of Worksop off the A616.

A statue of Robin Hood rears from the shrubbery at Thorseby Hall

TISSINGTON
Derbyshire

Set in the heart of the Derbyshire Peak District, little Tissington is a popular spot with visitors, and especially with that increasing number of travellers who arrive there on foot or bicycle down the Tissington Trail, which follows the route of the old railway from Buxton and is therefore very level, a boon to hikers and cyclists. The centre of Tissington is marked with a green, around which stand solidly built cottages backed by the village church and a fine Jacobean manor house.

Tissington is one of those villages where well-dressing is still practised on Ascension Day. The custom is said to date back to the time of the Black Death. The village wells are decorated with pictures made by pressing flower petals into frames filled with a layer of soft mud and moss. The effects are both lifelike and artistic, depicting religious or historic scenes — an attractive custom from the distant past.

TISSINGTON lies 6 miles north of Ashbourne, off the A515.

THE TRAMWAY MUSEUM
Derbyshire

The Tramway Museum at Crich contains the national collection of trams, some of which are still working, on a mile or so of track within the museum grounds. There are about 40 various vehicles, ranging from an 1874 horsedrawn tram from Sheffield to that city's last tram which carried its last passenger as recently as 1960. The Museum and Tram Sheds can be visited throughout the year, and provide the perfect place to visit on a rainy day. Other sights within the park are a monument to the Sherwood Foresters Regiment and a display devoted to the history of the Derbyshire lead mines.

CRICH lies 5 miles south of Matlock on the B5035.

THE WATERWAYS MUSEUM
Stoke Bruerne, Northamptonshire

This museum, set beside the Grand Union Canal, is one of the finest small museums in the country, devoted to recreating the life, times and history of the 200-year span of the Canal Age, before it was first threatened and then destroyed by the coming of the railways. There are narrow boats and all manner of interesting curios and artifacts. Ideal for children.

STOKE BRUERNE lies off the A43 road, 3 miles east of Towcester.

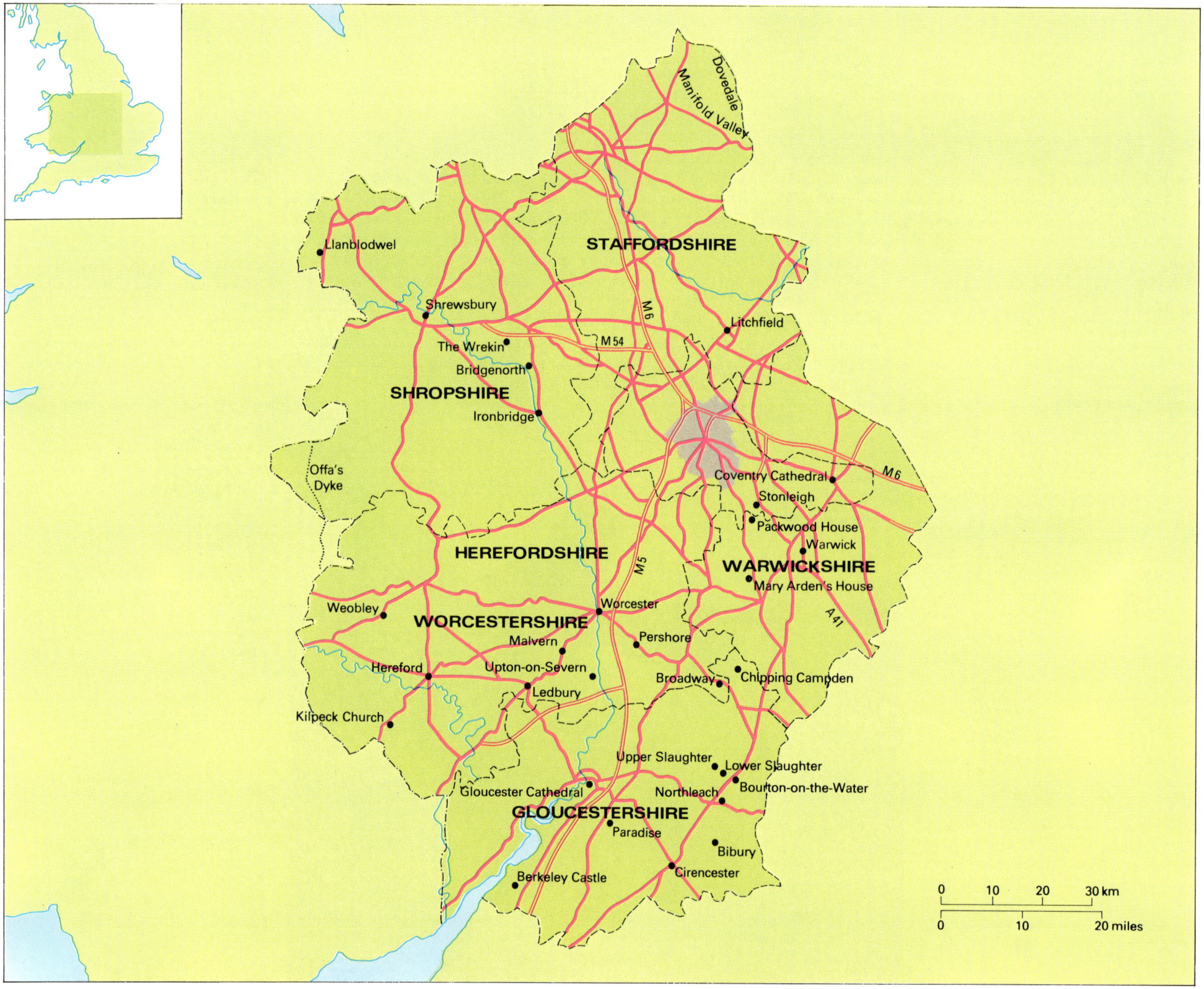

STAFFORDSHIRE
SHROPSHIRE
HEREFORDSHIRE
WORCESTERSHIRE
WARWICKSHIRE
GLOUCESTERSHIRE
Dovedale
Manifold Valley
Llanblodwel
Shrewsbury
Litchfield
M6
M54
The Wrekin
Bridgenorth
Ironbridge
Offa's Dyke
Coventry Cathedral
M6
Stonleigh
Packwood House
Warwick
M5
Mary Arden's House
Weobley
Worcester
A41
Malvern
Pershore
Hereford
Upton-on-Severn
Broadway
Chipping Campden
Ledbury
Kilpeck Church
Upper Slaughter
Lower Slaughter
Bourton-on-the-Water
Gloucester Cathedral
Northleach
Paradise
Bibury
Cirencester
Berkeley Castle
0 10 20 30 km
0 10 20 miles

The Heart of England

Set in a garden full of trim hedges, the home of Shakespeare's mother, Mary Arden, is an excellent example of a yeoman's dwelling

RUNNING ACROSS FROM the green and gold country of the Cotswold hills to the Marches of Wales, this part of England offers extremes. In the beginning, among those small, stone-built Cotswold villages, each tucked away in the valley, the air is soft, the landscapes gentle. The Cotswolds are almost archetypal England, a place which people visiting for the first time find very beautiful and somehow familiar. Places like the Slaughters, Bibury, Bourton-on-the-Water, or the glorious town of Chipping Campden are quite unmatched for beauty. Many of these Cotswold towns grew to wealth and prominence during the Middle Ages, when the wealth of England grew on the sheep's back. The country flourished by selling wool to the cloth looms of Flanders, and from the profits of this trade the burghers endowed their home towns with fine houses and magnificent churches.

Move west, out of these sheltering hills and the land changes, becomes more open, and the sky lifts, offering great sweeping views across the fruit orchards of the Vale of Evesham and across the hills to that old embattled frontier, the Welsh Marches. In this part of England, historical provincial cities have seen their share of strife, and record the events of more jarring days in their great castles, old walls, fine cathedrals — Berkeley Castle and Warwick are but two well-preserved examples of the medieval fortress — while few cathedrals in England can compete with Gloucester, Worcester or Hereford.

BERKELEY CASTLE
Gloucestershire

Even those who love castles cannot claim that Berkeley Castle is beautiful. But then it was not built for beauty but for utility and defence in times of war, and that original purpose shows. The walls are 14 ft thick, the Keep is Norman, the Great Hall dates from the 14th century, as do the kitchens. The castle is best remembered as the place where King Edward II was tormented, starved and finally murdered with a red-hot poker. Today, this castle, which has been a seat of the Berkeley family for over 700 years, is still in magnificent condition. The state apartments are full of armour, weapons, banners and paintings, while the terrace gardens and deer park are well worth seeing, as is the mainly Georgian Berkeley town itself, where, in the Norman Parish Church many of the Berkeleys lie in their tombs.

BERKELEY lies 20 miles north of Bristol off the A38.

BIBURY
Gloucestershire

Bibury is almost too pretty. It straggles up a hill above the River Coln, where a trout hatchery attests to the purity of the water, and is one of those places which gentrification has hardly touched. The cottages on Arlington row were hailed as just perfect by William Morris a century ago, and they still look right today. The Church of St Mary is basically Saxon, and Arlington Mill dates from around the 1650s. Bibury is not one of those places with one overriding attraction. Rather it is a place which looks just right in its setting and pleases all the senses with a fine touch all its own.

BIBURY lies on the A433 8 miles north-east of Cirencester.

BOURTON-ON-THE-WATER
Gloucestershire

Bourton-on-the-Water sounds as beautiful as it is. This is a large Cotswold village, full of attractions and popular with tourists. The River Windrush flows right through the village and is spanned at regular intervals by low bridges. The village is full of fine Georgian houses, there are various museums, curio shops, antique shops and a famous model village, also a fascinating aviary containing species of birds from all over the world — but do not let this modern touristic covering put you off, Bourton-on-the-Water.

BOURTON-ON-THE-WATER lies on the A429 3 miles south of Stow-on-the-Wold.

BRIDGNORTH
Shropshire

Like so many of these Shropshire towns, Bridgnorth stands looking out across the River Severn, keeping a wary eye on the Welsh. It is, of course, medieval, and has not grown much, or changed a lot since Tudor times, so it remains full of old houses, narrow streets, all clustered within what remains of the old ramparts. There are in fact two parts to the town, High and Low, one on the ridge, the other by the river, linked by a cliff railway and a series of steep steps hacked out in the limestone. It has been compared to one of those Tuscan *roccas* so common on the hilltops of Italy, and there are distinct similarities, even to the leaning tower of the ruined 12th-century castle and the warren of little streets. As one of the least-known towns in one of England's least-known counties, Bridgnorth is well worth a diversion.

BRIDGNORTH lies 20 miles south-east of Shrewsbury, on the A458, west of Wolverhampton

BROADWAY
Worcestershire

Broadway has been called the Show-Village of England, a place the tourist books like to present as typical of all English villages. If this were so, it would be no bad thing for Broadway is very attractive, but it can stand on its own as a fine example of a Cotswold village, and of traditional English country architecture. It is built in that soft yellow and white Cotswold stone that mellows so attractively, and with a setting of smooth lawns, cropped hedges and trees, the whole place is a picture. The High Street runs up from the wide Green where the Broadway Hotel and the famous Lygon Arms compete for attention, rising steadily to Fish Hill, where the ancient Fish Inn looks a little out of its depth. Carry on climbing, and you will arrive by Broadway Tower, a folly built in 1800 by the Earl of Coventry. From here you can see the spires of Worcester Cathedral, 20 miles away, but the best place to be on any warm summer evening is down in Broadway itself, enjoying the quiet corners of this pretty English village.

BROADWAY lies on the A44, 5 miles south-east of Evesham.

CHIPPING CAMPDEN
Gloucestershire

Chipping Campden is a wool town, one of those places that grew to prosperity off the sheep's back, for Cotswold wool was the main export of medieval England, a source of the country's riches. It is as a reminder of

A row of cottages built in Cotswold stone stand beside the mill stream at Arlington Row in Bibury

this fact that the Lord Chancellor of England sits on the Woolsack to this very day. Chipping Campden is a glorious town, full of fine buildings. The Woolstaple Hall was built in 1380, and parts of the Town Hall are even earlier. In the High Street stands the much photographed 17th-century Market Hall, a magnificent building, as is the medieval church, dedicated to St James and built by William Grevel 'the flower of all the wool merchants of all England', according to his memorial. This church is full of fascinating corners, glass, brasses, tombs, a wonderful find for any lover of the medieval world, but the same might be said of the whole town. Not much has changed here since the first Queen Elizabeth sat on the throne, and it remains a gem of old England, the perfect example of a market town.

CHIPPING CAMPDEN lies 10 miles north of Stow on the Wold, by the A424 and A44 roads.

CIRENCESTER
Gloucestershire

Unlike many of the other towns hereabouts, which grew up in the Middle Ages, Cirencester is Roman. Known then as *Corinium*, it was the second city of the country after London, and recent excavations have revealed just how extensive the Roman city was. The Roman walls still encircled the town in the Middle Ages, but the success of the wool trade can be seen here as elsewhere, most notably as always, in the magnificent 15th-century parish church of St John the Baptist, which is rightly regarded as one of the most perfect Perpendicular style churches in England. It is useless to describe St John's, for words are just inadequate; it has to be seen. Do not leave Cirencester without visiting the Corinium Museum, and Cirencester Park, which is now open to the public and set off by the great house built by Earl Bathurst in 1718.

CIRENCESTER is 19 miles south-east of Gloucester.

COVENTRY CATHEDRAL
Warwickshire

There is not much left of the old Saxon city of Coventry, where Lady Godiva rode through the streets on a white horse to spare the citizens from her husband's rapacious tax gatherers; the German bombers saw to that. On the night of 14 November 1940, Coventry endured one of the most sustained air attacks of the War and most of the city was totally destroyed, including the magnificent medieval Gothic Cathedral.

Modern Coventry is simply not attractive, but Sir Basil Spence's Cathedral, begun in 1954 and consecrated in 1962, is a modern marvel, and must be seen. The ruins of the old Cathedral act as a stepping stone to the new one, which is built in concrete and full of works of art. The porch contains a statue of St Michael by Epstein, the tapestry of Christ behind the High Altar is the largest in the world and was woven in France, the baptistry window is by John Piper and is a riot of colours, while the Chapel of Christ in Gethsemane contains a magnificent mosaic by Sykes. Perhaps the most effective memorial is a little iron cross forged from nails found twisted by fire in the ashes of the old Cathedral.

COVENTRY lies 12 miles north of Warwick on the A46.

DOVEDALE
Staffordshire

Though chiefly connected with Derbyshire, the Peak District laps over into Staffordshire, where Dovedale is

A view of Dovedale from Thors Cave, high above the river

one of the most beautiful of the dales. Dovedale is a lovely valley and the whole of it, from Hartingdon down to the stepping stones near Thrope, is a well-famed beauty spot, and rightly so. Two places to visit are Thors Cave and Weltonmill. Tors Cave is a strange place, the entrance alone being 60 ft high, set 200 ft or more above the river. The Dove actually marks the boundary between Derbyshire and Staffordshire, and is a famous fishing river, the happy hunting ground of Izaak Walton and his boon companion Charles Cotton. At any time of the year this deep little valley, with its river rushing fast along its bed, is one of the great natural attractions of this little-known county.

DOVEDALE lies east of the A515 between Ashbourne and Hartington.

The aisles of Gloucester Cathedral, lined with fine stained glass windows

GLOUCESTER CATHEDRAL
Gloucestershire

Gloucester is an industrial city today, but it does have one attraction that makes a visit here more than worthwhile; the great Gothic Cathedral, full of memorials, evocative of England's colourful history.

The cathedral towers over the workaday city and is dominated in its turn by a soaring Perpendicular tower, which was erected in the mid 15th century. The cathedral nave is Norman, and stands on the site of a Saxon abbey with a 14th-century trancept, a 15th-century Lady Chapel and medieval cloisters to top off the work. The interior glows with colour as light streams through the huge 72 ft-high East Window — the largest stained glass window in Britain — and illuminates the old battle flags, stonework, brasses and tombs. Most notable of these is the 14th-century tomb of Edward II who was murdered in nearby Berkeley Castle. Gloucester Cathedral is one of the centres for the tri-annual Three Choirs Festival.

GLOUCESTER lies 9 miles west of Cheltenham, on the A40, or M5 motorway.

HEREFORD
Herefordshire

Hereford is an unusual market town, for while many are attractive, if a touch bucolic, Hereford has a long connection with the arts. Garrick the playwright was born here; Kemble and Sarah Siddon came to play here; and the town is one of the founder cities for the famous Three Choirs Festival, which dates back to 1720.

Going back even further, Hereford is a Border town, a bastion on the Marches of Wales. Parts of the old walls still remain, there are many attractive half-timbered houses to beautify the old streets and the cathedral is said by cathedral lovers to be one of the finest in the kingdom. It is a very early Norman building, with a 14th-century central tower, and full of all the best cathedral features, including stained glass, brasses, tombs, fine vaulting, chantry chapels, the lot. There is also a notable chained library and a very old map of the world. Hereford is also a good centre for touring north along the Marches, or south to the Severn Valley.

HEREFORD lies 13 miles south of Leominster on the A49.

IRONBRIDGE
Shropshire

Perched on limestone cliffs above the River Severn, the gorge spanned by that famous iron bridge that gives the town its name, this town set the benchmark for the Industrial Revolution that transformed England from a largely agricultural community into an industrial society. Today, the town is a great centre for the study of

A view of Ironbridge where the Industrial Revolution really began

industrial archaeology, with several museums recording the history and development of the local iron industry. After visiting there though, remember to go closer for a look at the great iron bridge itself. It was made in 1777, the first bridge built in iron in England. Now take an even closer look and you will see that while the metal is iron, the method was carpentry — note the joints of the beams, the slotting of the cross-members, exactly as used on wooden bridges. The bridge still arches high above the river, a magnificent sight. Not far away, in the suburb of Coalbrookdale, the ironmaster Abraham Darby first discovered how to smelt iron with coking coal, and although this is probably the birthplace of the Industrial Revolution, it remains quite a pretty place, with a number of half-timbered houses. England's history is by no means all medieval, as Ironbridge proves.

IRONBRIDGE lies 6 miles south of Telford on the A4169.

KILPECK CHURCH
Herefordshire

Kilpeck Church is most attractive, one of the best preserved Norman churches in England. There is a huge Norman font and some magnificent Norman carving on the south door. It stands beside the remains of a Norman motte, and probably dates from about 1135. The style which we call Norman is known on the Continent as Romanesque, and there are distinct similarities between the rich carvings

Rich Romanesque carvings decorate the doorway of Kilpeck Church

at Kilpeck and those found on the great pilgrimage churches of south-west France, on the famous Road to Compostela, tempting speculation that the master-mason who worked at Kilpeck had followed that road to the Shrine of St James at Santiago de Compostela in Galicia.

KILPECK lies 6 miles south-west of Hereford, off the A465.

LEDBURY
Herefordshire

Set beneath the bare green slope of the Malvern Hills, Ledbury is a very attractive market town which contains a considerable number of those pleasing black and white half-timbered houses seen in this part of England. Two worth noting are the Old Market House, which has been hoisted onto wooden pillars in the High Street, and the much restored Grammar School in Church Lane, where a museum charts the development of Ledbury from Anglo-Saxon times. Those who like visiting old churches will enjoy the medieval Church of St Michael and All Angels, though the spire dates from as recently as the 1730s. The interior contains a good number of brasses from the 14th to the 19th centuries.

LEDBURY lies 8 miles south of Malvern on the A449.

LITCHFIELD
Staffordshire

Litchfield is best known to the world as the birthplace of that 18th-century writer and sage, Dr Samuel Johnston, who was born here in 1709. His birthplace is now a museum and his statue stands in the market place. The town dates back to Roman times, although it is the Georgian facades that strike the eye most pleasantly today, as well as the fine cathedral

Three tall spires set off the rooftops of St Chad's Cathedral built between 1195 and 1235

dedicated to that little-known cleric, St Chad, Bishop of Mercia about AD 660. The pilgrimage to his shrine raised the money for this huge three-spired cathedral. It was built between 1195 and 1235 and can be seen for miles, but the west front repays closer inspection for it is covered with statues and carvings. The Lady Chapel contains some fine 16th-century glass and the Cathedral treasury holds St Chad's Gospels, a rare collection of early Gospels. Taken all in all, Litchfield is one of those towns that all true travellers should visit.

LITCHFIELD lies 20 miles south of Stafford on the A51.

LLANBLODWEL
Shropshire

Llanblodwel sounds very Welsh, but it is an English town, one of those graceful little places on the eastern side of the Marches. It goes back to the early days of the 12th century, when the Normans first arrived at this far frontier, and built the Church of St Michael the Archangel, which after 800 years has evolved into an amazing mixture of architectural styles, with a bit of this and a bit of that, a Norman doorway, a detached bell tower built in the last century, a spire set on top like a witch's hat, a marvellous carved south porch. There is a medieval bridge over the river, and a 16th-century inn.

LLANBLODWEL lies off the A495, 4 miles south of Oswestry.

MALVERN
Worcestershire

The two boroughs of Malvern, Great Malvern and Malvern Link lie on the side of the Malvern Hills, a curious,

A view of the green and winding Manifold Valley, one of the less well-known beauty spots of Derbyshire

high, nine-mile long ridge that dominates the countryside round about. In the last century Malvern developed into a spa, thanks to the purity and abundance of Malvern Water, which is still bottled here today. This era has left Malvern with a fine collection of hotels, and it now acts as a tourist centre, lying as it does in some splendid countryside, with first-class walks on the hills above. Great Malvern also contains a marvellous relic in the shape of the Priory Church founded during the reign of William the Conqueror, much of which survives. The carving on the choir pews shows the labours of the seasons, the stained glass is magnificent, and the building practically unique. For this and all the other reasons given here, Malvern should not be missed on any journey through this region of England.

GREAT MALVERN lies 10 miles south of Worcester, on the A449.

THE MANIFOLD VALLEY
Staffordshire

The River Manifold is a tributary of the River Dove, running into the larger river through the Staffordshire Peak. It rises, like the Dove, close to the Traveller's Rest Inn on the Buxton Road near Leek, and the upper reaches of this river are quite spectacular, the river cutting a deep gorge in the gritstone rock, and after winding down a deep limestone valley, the Manifold suddenly narrows at Eaton Hill and finally flows into the Dover near Ilam. Sights to see on the way include Darfor Bridge and the place where the river vanishes underground for a while near the Redhurst Swallets, only to reappear in the grounds of Ilam Hall, a National Trust property where the river's reappearance can be seen. The name Manifold refers to the meandering way it winds across the country — there is no industrial connection.

THE MANIFOLD VALLEY and river can best be viewed near Hulme End, north of the A523.

An inside view of Mary Arden's house at Wilmcote

MARY ARDEN'S HOUSE
Warwickshire

No tour of Warwickshire would be complete without some reference to the county's most famous son, William Shakespeare of Stratford-on-Avon. Indeed, memorials and reminders of Shakespeare abound in the county, but Wilmcote was the home of his mother, Mary Arden, and her house can still be visited. It is a rather fine half-timbered building of the classic Tudor style, so she was clearly a lady of property; her father was in fact a wealthy yeoman. She married John Shakespeare of Stratford in 1557, and moved into town, but her house at Wilmcote is now a Shakespeare Museum.

WILMCOTE lies off the A34, 3 miles north-west of Stratford.

NORTHLEACH
Gloucestershire

The Cotswolds are full of attractive villages, many of them well established on the tourist trail, but little Northleach has been overlooked. It lies in the valley of the River Leach, just north of the A40 road, and is noted in particular for the Church of Sts Peter and Paul, a very large 15th-century building, built, like so many of these impressive Cotswold churches, from the profits of the local wool trade. The west tower is lofty, and the porch is full of statues is lofty, and the porch is full of statues and delicate tracery. Inside there is a remarkable brass to John Fortey, who built the clerestory in the Chancel, and an equally remarkable font. All in all, this church contains some of the finest elements of English Gothic.

NORTHLEACH is 10 miles north east of Circencester on the A429.

OFFA'S DYKE
Herefordshire

No county can really lay claim to Offa's Dyke, a vast ditch and earthwork constructed in the 8th century by command of King Offa of Mercia. It runs for nearly 200 miles along the Marches of Wales, from the Dee in the North to the banks of the River Wye near Chepstow.

It cannot have been Offa's intention to defend this land, and the opinion of modern historians is that he had it built to mark the frontier between his kingdom and the Welsh. The Dyke today provides the basic route for the Offa's Dyke Footpath, a 168-mile journey over countless stiles, a challenge indeed to the fit, well-equipped hill walker.

OFFA'S DYKE can be followed from various points along the Welsh border, notably near Hay-on-Wye.

Morris Dancing

Morris Dancing is a uniquely English custom. It dates back certainly to the early medieval period, but the origin of this custom is probably pagan, when the people danced to celebrate the start of spring or to ensure the fertility of their crops. It has to be said also that while Morris Dancing is often done for the delight of visitors and tourists, it is part of a genuine local folklore, and the people would still dance whether there were visitors to watch or not.

Morris Dancing can be found in Yorkshire, in the Midlands or many parts of the West Country, often in a distinctly different local form, or with additions and embellishments unique to a particular village. The Cotswolds, however, maintain a regional form of this dance, the almost classic 'Cotswold-Morris' which can be seen at its best on any summer Bank Holiday Monday, in the village of Bampton.

Cotswold-Morris Dancers are all men, and the Bampton team starts dancing soon after daybreak, and continues to dance throughout the day, although with frequent pauses for rest and refreshment in the various local pubs. The dance ends officially with tea on the Vicarage lawn, but if the dancers are still in the mood, as they always are, it then continues well into the evening. The music is provided by fiddle, accordian or tambourine, helped out by the Morris bells lashed to every dancer's leg below the knee, and while every dancer is dressed in white, some have their faces black, which gives rise to the theory that the word 'Morris' is a version of 'Moorish', which seems somewhat unlikely in this typically English setting.

Morris dancing by the Compden Morris Men at Chipping Camden in the Cotswolds

Fine 18th-century wrought-ironwork enhances this balcony in a home in Bridge Street, Pershore

PACKWOOD HOUSE
Warwickshire

Packwood House is an elegant Tudor manor house built about 1550, and much expanded about 100 years later. Pleasant though it is, Packwood's main attraction for visitors is the garden, and the reason it came to be constructed. When the Civil War broke out in 1642, the then owner of Packwood, John Fetterston, could not decide between King and Parliament, so he decided most wisely to stay at home. He collected sundials, of which several still remain in the grounds, and was an expert at topiary, so during the long years of the War he clipped and trimmed the yew trees of Packwood to represent the Sermon on the Mount. The house contains needlework and tapestry, but the gardens should not be missed.

PACKWOOD HOUSE lies at Lapworth, near Knole, off the A34 south of Birmingham.

PARADISE
Gloucestershire

It would be a very world-weary traveller indeed, who could resist visiting a village called Paradise. It is really little more than a hamlet, lying just off the A46 road, north of Stroud, and it had no name until King Charles stumbled upon it during the Civil War. On hearing that this beautiful little place had no name, he told them to 'call it Paradise'. The setting by the little River Slad is quite delightful, and it makes a pleasant stop on the way to the Benedictine Abbey at Prinknash, which may be visited, together with its working pottery and craft shop.

PARADISE lies off the A46, 3 miles north of Painswick.

PERSHORE
Worcestershire

Pershore is not one of those towns that many English people think about or tourists visit, and this is passing strange because it has a lot to offer. It lies on the River Avon on the western edge of the Vale of Evesham, so the setting alone is very beautiful. Penetrate to the centre and you will discover an almost intact 18th-century Georgian town, with a square and High Street lined with row after row of elegant buildings, most pleasing to the eye. Pershore Abbey, which was the reason that the town grew up here, was destroyed during the Reformation, but the Abbey Church was preserved to serve the parish, as it still does. From here it is just a short step to the delicate footbridge across the river, which gives visitors a fine view of the old houses on the banks. On any journey into Wales, a diversion through Pershore could be most rewarding, as trips off the too-well-trodden track usually are.

PERSHORE l.es 6 miles west of Evesham on the A44.

SHREWSBURY
Shropshire

Set on a great winding loop of the River Severn, one of the Border fortresses guarding the Welsh Marches in days gone by, Shrewsbury has somehow managed to preserve all that is best of the old town through much recent reconstruction — in short, it still looks right.

It takes the visitor a day or two to get to grips with Shrewsbury, for it is a narrow, tightly-knitted town, a place of narrow streets and old houses, with ruins and churches. The castle is Norman and the gateway is still exactly as it was 500 years ago, even if the rest has been over-restored. St Mary's Church is also Norman, with plenty of later additions including some fine glass, and there are plenty of monastic buildings around the town, though Shrewsbury School dates from the 16th century, after the Reformation. The whole town is worth exploring, from the banks of the Severn to the castle up above, as a splendid relic of the High Middle Age, and a very fine city of today.

SHREWSBURY lies 11 miles west of Wellington by the A5.

THE SLAUGHTERS
Gloucestershire

The twin villages of Upper and Lower Slaughter, set beside the little River Eye, in a green fold of the Cotswolds, are quite delightful. This is a place of mellow Cotswold stone, of ivy-draped drystone walls, of birdsong and the tinkle of water in the brook, everything an English village — or villages — should be. The two villages are a mile or so apart, and most of the buildings date from the 16th or 17th centuries. There is a fine 16th-century dovecote in the grounds of the Manor Hotel in Lower Slaughter, while the Church of St Mary contains many memorials to the local squires, the Whitmores. The centrepiece of this village is The Square, which has lots of old cottages and the old corn mill which still retains the water wheel, and there are more of these attractive golden stone houses on the banks of the Eye.

Upper Slaughter, as the name implies, stands on a hill. The remains of a Norman motte and bailey (hill and ditch) castle still stand behind Home Farm, but the name 'Slaughter' has nothing to do with any medieval massacre. It is from the Saxon 'sloutre' — which gives us 'Slough' — a muddy place, and it can get very muddy indeed hereabouts, especially during the winter months.

THE SLAUGHTERS lie 3 miles west of Stow on the Wold, off the A436.

A view across the green and stream into the centre of Lower Slaughter, one of the most attractive Cotswold villages

STONLEIGH
Warwickshire

Stonleigh can rightly claim to be one of the loveliest villages in Warwickshire, that country of lovely villages. It is not very large, but what there is, is choice, from the Tudor almshouses to the 14th-century bridges over the river, and the solid Norman church. The village has belonged for centuries to the Leighs of Stonleigh Abbey. Alice Leigh built the almshouses here in 1594, two years before she married Robert Dudley. She lived on until well into the next century, supported the Royalists during the Civil War and was made a Duchess by Charles I in 1645. Stonleigh is a place to wander about in, crossing the River Sowe by Sowe Bridge, looking at yet more of those beautiful black and white houses.

STONLEIGH lies 3 miles east of Kenilworth, off the A46.

UPTON-ON-SEVERN
Worcestershire

Upton-on-Severn is not very large but it has a long and fascinating history, and a good range of architecture. It lies on the west bank of the River Severn, overlooked by the green Malvern hills on one side, and by Bredon Hill on the other. It is now an important boating centre for the Severn valley, with a large marina and plenty of moorings. A number of good riverside pubs and a series of coaching inns line the main street. Clearly, despite its present obscurity, little Upton was once a place of importance. It was a strategic centre during the Civil War and the Upton Heritage Centre, in the old tower known as the Pepperpot, recounts the story of those Cavalier-v-Roundhead battles fought on this soil long ago.

UPTON-ON-SEVERN is on the A4101, 12 miles south of Worcester.

WARWICK
Warwickshire

By any standards, Warwick is a splendid place. It may look like just another pretty market town, but Warwick is more than it seems, for this is, or was, the home of the mighty Nevilles, Earls of Warwick, makers and breakers of kings.

There is so much to see here that it is hard to know where to start, but the castle must dominate any visit, and is clearly the glory of the town. Parts of it date from AD 900, but most of the present buildings were erected during the jarring days of the 14th and 15th centuries. Caesar's Tower and Guy's Tower overlooking the Avon, are filled with treasures, armour, paintings, tapestries, furniture, the collections of that great series of families, Beauchamp, Neville and Dudley, who held the earldom of Warwick and their symbol, the bear and ragged staff, can be seen everywhere in the town. In Warwick Castle, Piers Gaveston was tried and condemned, and here Warwick the Kingmaker pondered on who to support during the Wars of the Roses. If this is too much, then visit a still surviving relic at the Lord Leycester Hospital, now a home for old soldiers and endowed by Robert Dudley, Earl of Leicester and favourite of Queen Elizabeth I. Pass on from there to explore the streets of this historic town, to see especially the Beauchamp Chapel in the Church of St Mary, which contains the tomb of its founder, Richard Beauchamp, Constable of France to Henry VI, and those of many other Earls.

WARWICK lies 8 miles north of Stratford-on-Avon by the A46.

The Wrekin, a round green bump of a hill crowned by an Iron Age fort

WEOBLEY
Herefordshire

Weobley — pronounced Webley — is one of the famous black and white villages which are not uncommon in this region of England. It can be said though, that Weobley is one of the finest, a place where striking timber-framed houses fan out from the main street, and in the narrow byways beneath the sharp spire of the church. A local man, Ben Tomkins, is credited with breeding those black and white Hereford cattle back in the 18th century, perhaps to match the architecture. The Church of Sts Peter and Paul dates from the 14th century, and contains among other memorabilia, a fine statue of Colonel John Birch, a Cromwellian officer who regularly fell out with Cromwell and was imprisoned on countless occasions. Weobley is a place that can be found only in England, and then only on the March of Wales, and it alone would make a trip there worthwhile.

WEOBLEY lies on the A4112, 12 miles north-west of Hereford.

WORCESTER
Worcestershire

Worcester is a very fine, very historic city on the River Severn. It has been an important place since Celtic times, and seen a lot of strife down the centuries, yet somehow preserved some fine buildings, and a rather peaceful air. The town is built on both banks of the river, and as the Severn is one of England's less predictable streams, the city gets flooded from time to time. One place unravaged by fire or flood, is Worcester Cathedral. Parts of this date from the 7th century, but it has been added to, or rebuilt, many times, and the pleasing whole we have today is on that account rather remarkable. It took 200 years to complete the Nave, which contains the tomb of King John, who loved the place so much that he wanted to be buried here rather than at Westminster or the Plantagenet mausoleum at Fontevraud. The old Bishop's Palace at Hartleberry contains the County Museum, and has an impressive collection of gypsy caravans.

WORCESTER lies 12 miles north-west of Pershore, on the A44.

THE WREKIN
Shropshire

The Wrekin is a beautiful hill, very distinctive, lying a little to the south of Wellington. On the summit, stand the smooth grass ramparts of an Iron Age hill fort and those who plod to the top will be rewarded with wonderful views over a broad band of country, and looking south, see the high, green and russet ridge of Wenlock Edge. The hill fort was probably the marshalling place for the warlike Cornovi tribe of Britons who lived here on the wild march for centuries before the Romans came to Britain, 2000 years ago.

THE WREKIN is just south of Wellington, which is north west of Telford on the A442.

YORKSHIRE
LANCASHIRE
HUMBERSIDE
CHESHIRE
IRISH SEA
Richmond
Wensleydale
Jervaulx Abbey
Ripon
Fountains Abbey
Ripley Castle
Harewood House
Whitby
Hulton-le-Hole
Pickering
Scarborough
Filey
Flamborough Head
Bridlington
Sledmore House
York
Burton Constable Hall
Howden
Epworth
Settle
Skipton
Morecambe Bay
Lancaster
Blackpool
Astley Hall
Port Sunlight
Speke Hall
Chester
A1
A19
A64
A166
A165
A65
A59
M6
M55
M61
M58
M62
A627(M)
M1
M18
M180
M53
M56
0 10 20 30 km
0 10 20 miles

Yorkshire, Humberside & North west

Sheep crop the grass on Lewisham Moor in Yorkshire, a patchwork of green fields stretching as far as the eye can see

YORKSHIRE IS, WITHOUT DOUBT, one of the finest of all the English counties; for it combines a striking mixture of neat farms, green dales, high hills, tumbling waterfalls, open windswept moorlands, an attractive coastline and a benevolent climate with ancient cities, strong castles, soaring cathedrals, and the usual pleasing pattern of small villages, little market towns, and here and there among the green valleys, the jumbled old stone litter of some ancient ruined abbey.

Not only the abbeys have passed, but Yorkshire itself has also changed, with the ancient 'Ridings', North, South and West now changed into sub-divisions and newly-created counties, South, West and North Yorkshire which with Humberside now blankets the old county which was, in its day, the largest county in England and held, in the City of York, the second town of the Kingdom. All that may have been swept away in the county reorganisation of the 1970s, but nothing else has changed and Yorkshire remains magnificent, a beautiful, varied county.

Yorkshire is the traveller's county whether you ride about by car or bicycle, or on one of the famous railway lines like the Settle to Carlisle or the North Yorkshire Moors Railway.

Next door Lancashire is an underrated county, best known perhaps for Blackpool, the Jewel of the North, with its Golden Mile, its 518 ft tower, its Victorian-style entertainment and ultra-modern facilities.

ASTLEY HALL
Lancashire

Astley Hall is a curious place, to say the least. It was built, or rather begun, in Elizabethan times, so that the basic structure is Tudor. This classic, central core is however totally overshadowed by the 17th-century additions of the great many-windowed bays which flank the main building, so that at first sight, Astley Hall glitters with the reflections from the glass.

Once inside, visitors can inspect the great bed, called Cromwell's Bed, because the Lord Protector is said to have slept in it before the Battle of Preston. Note the extensive dining table in the Long Gallery.

ASTLEY HALL is in Park Road, Chorley — Chorley lies just south of Preston on the A6.

BLACKPOOL
Lancashire

England has two great and vastly popular seaside towns, Brighton in the South and, on the coast of industrial Lancashire, Blackpool.

Of the two, Blackpool is the larger, the more colourful, the less elegant, but the most fun, and as a seaside town in the best sense of the word, offers all the elements of the English seaside holiday. There are donkeys on the sands, bandstands, dancing in the Tower ballroom where the organ plays, all the sweetshops and stalls sell Blackpool rock, the illuminations are famous and draw the crowds by the coach-load every year, and the famous Blackpool Tower looms over it all, a landmark for miles around.

Out of season, Blackpool is much quieter, and then the conference season begins, attracting vast numbers of more serious folk to enjoy the sights before summer comes again, but those who enjoy a little fun but prefer the quiet life can always play golf around Lytham St Anne's or just take a walk along the beach early in the day before the tourists arrive.

BLACKPOOL lies 15 miles west of Preston on the A583.

Blackpool's colourful illuminations

BRIDLINGTON
Humberside

Bridlington, set in the sheltered southern side of Flamborough Head, is first and foremost a fishing port and only incidentally a tourist centre, although the fishing has declined and the tourist trade flourishes. Like a number of the old Yorkshire towns, it was once a pilgrimage centre, and had a famous Priory, founded in 1114 and greatly extended throughout the Middle Ages. John of Bridlington, the local saint, was canonised in 1401, and although the Priory went the way of most monastic buildings in 1538, the Priory Church is now the Parish Church of St Mary, still a glorious building, though over-restored by the Victorians. The Priory Gatehouse now contains a museum, but those who have no love of history or architecture will find plenty of interesting walks around the old fishing port, on the sandy beaches or down by the harbour.

BRIDLINGTON lies 20 miles south of Scarborough on the A165.

BURTON CONSTABLE HALL
Humberside

The masters of their craft worked to beautify Burton Constable Hall and make it the place we see today. They were helped by the Tudor mansion on

The rows at Chester, black and white 'magpie' style architecture

which it is based; this was built in 1570 and then altered by Robert Adam and Wyatt in the 1750s. They added another storey to the eastern side, and balanced this with the central block and pediment, while retaining the Tudor minstrel gallery and adding a Georgian dining room to the interior. The grounds and park were in the hands of Capability Brown, who cannot have visualised the attraction which exists there today — there is also a zoo and a model railway. Burton Constable is a fine house, and provides an interesting visit during the summer when it is open to visitors.

BURTON CONSTABLE HALL lies 1 mile north of Sproatley, off the B1238, east of Kingston-upon-Hull.

CHESTER
Cheshire

Chester is an ancient city dating back to Roman times, when it was founded as *Deva*, a base for the XXth Legion. It stands on a strategic site on the River Dee, close to the March of Wales, and remains an important commercial city, with several sights to attract the attention of the passing traveller, not least the Roman remains, the cathedral, the encircling city wall which provides the ideal route around the city and, of course, the famous 'Rows'. The Rows is a series of multi-storey shops and houses which line the main streets of Chester and are a unique feature of English architecture. Legend has it that the 'Rows' dates back to 1485, the year of Bosworth Field, but those which remain today are from the later Tudor and Stuart periods. They are a most unusual sight, running down Eastgate, Bridge Street and Watergate; clearly they are designed to provide good, covered shopping in the rainy north-west, but since they are both functional and especially attractive, one can only wonder why they have not appeared in other English cities. As it is they remain unique to Chester and this is one reason for visiting this elegant provincial city.

The cathedral dates back to the immediate post-Conquest period and stands on the site of a Saxon shrine. It began life in 1092 as the church of a Benedictine community, although it has been greatly enlarged and beautified down the centuries. The Treasury contains a fine collection of medieval manuscripts, while the nave has superb 14th and 15th-century stonework, including the gargoyle known as 'The Chester Imp'. Things to see in the countryside round about include Little Moreton Hall, perhaps the finest Tudor style mansion in Britain.

CHESTER lies 17 miles south of Liverpool.

EPWORTH
Humberside

Epworth really ought to be a famous spot, even though it is just a pretty little place tucked away in Humberside's countryside west of Doncaster. As it is, Epworth is hardly known outside the Methodist community, for John Wesley, the founder of Methodism, was born here in 1710, three years after his brother Charles, the author of some stirring hymns.

In fact, the building which claims to be Wesley's birthplace is actually the one built to replace the Wesley home which was burned down by a mob some years after John Wesley was born — and the motive was political not religious. It was the family home for several years, and is today a Wesley Museum, full of memorabilia and open during the summer months.

EPWORTH lies 2 miles south of the M180 motorway, west of Doncaster.

FILEY
North Yorkshire

Filey is one of the most popular seaside resorts of the Yorkshire coast, and blessed with a beautiful natural setting. The sands are wide, safe and gently shelving, well supplied with cliffs, caves and rockpools where crabs lurk to be chased out by small children armed with shrimping nets. There is something almost Edwardian about Filey, a place evocative of those childhood seaside holidays in times gone by, with walks along the cliffs, putting greens, golf links, concert parties and bandstands.

Old Filey, the fishing quarter of the town that existed here long before tourism was invented, is a quaint place, full of old houses and little courtyards called 'yards'. Charlotte Brontë often stayed in Cliff House in Bell Street, which is now a restaurant, and she mentions Filey frequently in *Jane Eyre*.

FILEY lies 10 miles south of Scarborough off the A165.

FLAMBOROUGH HEAD
Humberside

Flamborough is one of the great sea promontories of England, a vast, soaring cliff, rising up out of the grey

Flamborough Head lies in the sea like a piece of a giant's jigsaw puzzle

stormy waters of the North Sea. This is the place to go when an easterly gale is sweeping in across the sea from Norway and great waves send the spray soaring high into the air in sheets of foam. On less dramatic days there are seabirds by the thousand, an old lighthouse, and Flamborough Church, basically Norman. Andrew Marvell was the rector here.

FLAMBOROUGH HEAD lies 17 miles south of Scarborough.

FOUNTAINS ABBEY
North Yorkshire

In its heyday, during the 15th century, Fountains was the richest Cistercian abbey in England. The Cistercians preferred these quiet, remote locations for their monasteries and the community worked hard to transform this bleak valley into fertile farming country. The abbey dates from 1150, and although much was destroyed or carried away at the Dissolution, a considerable amount remains, notably the secular buildings, kitchens, dormitories and cellars, as well as the nave of the Abbey Church and the Chapel of the Nine Altars. The ruins and the gardens were landscaped in the last century and there is no more beautiful spot in Yorkshire when the sun is setting and the gardens are filled with the song of the birds and the sound of rippling water. Fountains Hall nearby, was built with dressed stone filched from these ruins.

FOUNTAINS ABBEY lies 4 miles south of Ripon on the B6269.

GARSDALE
Lancashire

Garsdale lies in that part of Lancashire which juts up into Cumbria and belongs more to those dales more commonly found in Yorkshire. It is a quiet, beautiful place, where the steep sides of the surrounding fells fall almost sheer to the valley floor where a rushing beck tumbles down towards the River Eden further to the west. This is a place to park the car and climb up some steep, winding footpath to the breezy fells.

GARSDALE lies west of Garsdale Head on the A684.

HAREWOOD HOUSE
West Yorkshire

Lying just off the main road between Leeds and Harrogate, Harewood House is just one of those great country houses which do so much to enhance the Yorkshire landscape. It stands in the middle of a vast park, and was built between 1759 and 1771 by a local builder, Richard Carr, although Robert Adam was charged with the interior decoration. Harewood House remains the finest example in Britain of the Adam style. The present Earl of Harewood has added a tropical aviary by the lake to an already long list of attractions which include a fine silver collection, a tea service once belonging to Marie Antoinette, and some magnificent Italian paintings. Harewood House is open daily in the summer and on a number of days during the winter months.

HAREWOOD village lies 5 miles south of Wetherby on the A659.

THE HERRIOT COUNTRY
North Yorkshire

When James Herriot wrote his first book, *All Creatures Great and Small,* he can have had no idea what it would do to much of Yorkshire, transforming Swaledale and a large

The magnificent gallery inside Harewood House

part of the county between Coverdale and the North Yorkshire Moors into 'The Herriot Country'. According to the author his fictional village of Darrowby is a combination of Thirsk, Richmond and Middleham — and a lot of imagination.

Even so, the Herriot Country does exist, and can be visited, and travellers will find it by wandering about along Swaledale, across the Vale of York to Middleham, north and west of York and south even as far as Selby, across dales and moorlands, a delightful and (with the books beside you) an evocative journey.

THE HERRIOT COUNTRY all lies within 50 miles of York, to the west and north.

HOWDEN
Humberside

Howden isn't very big but it has two features which make it outstanding. The first is the Horse Fair which Dickens wrote about, which is still one of the largest in the country. The second, and a more constant attraction, is the Church of St Peter in the centre, which seems far too large and magnificent for any possible congregation. The main part is medieval, with a fine 14th-century west front, but the interior woodwork is modern and bears here and there the mouse emblem of Robert Thompson. St Peter's is a very fine English church, topped by a bell tower.

HOWDEN lies 4 miles north of Goole on the A614.

HULTON-LE-HOLE
North Yorkshire

Hulton-le-Hole is a small, straggling village, lying a little to the north of the A170 and acting as a gateway to the Cleveland Hills and the North Yorkshire Moors. The houses lie on green banks on either side of Hatton Beck which plunges through this valley from the high moorland to the north, past Roseland Abbey which lies five miles away. Hulton was listed in the Domesday Book but most of the present village dates from the 17th and 18th centuries. The Rydale Folk Museum, next to the Crown Inn, is well worth a visit, for although small it presents a comprehensive picture of Yorkshire village life.

There are several pleasant pubs and a wide green, ideal for picnics beside the stream, while the other villages round about provide good objectives for walks or car excursions.

HULTON-LE-HOLE lies north of Kirkby-Moorside, off the A170, 7 miles north-west of Pickering.

JERVAULX ABBEY
North Yorkshire

Jervaulx is one of the great ruins of Yorkshire, an example — and a terrible one — of all that was lost when Henry VIII smote the monastic orders in the 16th century. Jervaulx was founded in 1155 and rose to become one of the most important Northern abbeys, a place of work, worship and learning unrivalled in the country, and noted for the quality of the sheep farming. The King's Commissioners were particularly brutal at Jervaulx, sacking the abbey, dispersing the monks and encouraging local people to level the buildings and carry away the stones.

JERVAULX ABBEY lies 2 miles south of East Witton on the A6108.

LANCASTER
Lancashire

Ask a hundred people to name six historic English cities and the chances

A quiet moment at popular Morecambe Bay

are that not one will mention Lancaster, and yet few places can match this historic county town for relics of the past. The Romans had a marching camp here and the town has remained an important trading centre ever since, but the historic heart of the town is the great castle built by the Normans, rebuilt by John of Gaunt and still in service during the reign of Elizabeth I.

Unlike other English castles, Lancaster has little romance. This one was built for war and oppression, and it shows in every stone. It records, in various collections, the savage laws of England since Magna Carta, and displays include instruments of torture, branding irons, the old gallows, the cells, and the prisoners' chapel.

LANCASTER lies 22 miles north of Preston off the M6.

MORECAMBE BAY
Lancashire

Morecambe Bay is one of the wonders of the North West, a vast area, drained twice a day by a huge tidal range which reveals mile after mile of sand flat and shallows, the sea flooding back to reflect the slow descent of the western sun in the evening.

The bay is surrounded by the attractive coastline of Lancashire and Cumbria, reaching from Morecambe itself all the way north and west to Grange-over-Sands and Walney Island on the tip of the Furness peninsula. The bay is fed by several rivers, the Kent and the Duddon in particular, which widen into estuaries as they pour into the bay. It is possible to walk across the sands between Grange and Morecambe, but the sands are dangerous, the tides rapid and this walk should only be made with an experienced local guide.

MORECAMBE lies 29 miles north of Blackpool.

Horn-blowing ceremonies

Yorkshire retains a number of ancient ceremonies, enshrined for all time in the customs of the county.

Two which can be seen regularly, and often every day for at least part of the year, are the Horn-blowing Ceremonies of Ripon and Bainbridge.

The horn-blowing in Ripon goes back well into the Middle Ages, to the time when the chief lay-officer of the city was known not as the Mayor but as the Wakeman. Among the Wakeman's daily tasks was to sound a horn at dusk as a signal to close the town gates and announce that the city had passed into his care for the night.

This custom continues to this day and can be observed every evening at 9 o'clock in the Market Square, when a horn-blower, wearing a tri-corn hat, sounds the horn four times, from each corner of the central obelisk. This ceremony is also recorded in the city motto 'Except the Lord keep ye Cittie, ye wakeman waketh in vain'.

The horn-blowing at Bainbridge, which takes place every evening in winter between Holyrood (27 September) and the evening of Shrove Tuesday, dates back to Norman times, when the forests hereabouts were the hunting grounds of the Lords of Middleham. The original intention was that the sound of the horn would draw benighted travellers through the outlaw-infested woods to the shelter of the city, and the sound can be heard up to three miles away.

Horn blowing at Ripon, Yorkshire

PICKERING
North Yorkshire

The railway station is, for most visitors anyway, the most important part of Pickering. Not that the town is unattractive, far from it, for it is a busy place and a market town for the surrounding district. Not much remains of the castle except some fragments of the old walls and keep, but many kings stayed here during the Middle Ages including King John and Richard II. The Parish Church is Norman and certainly well worth looking at, for it contains a remarkable array of wall paintings showing the lives of the apostles and dating from 1450. They are so fresh because they were covered during Puritan times and only uncovered again 100 years ago.

Having seen all that, and explored Pickering's excellent shops, there remains the railway station. This is quite old — at least 150 years — and is the terminal for that wonderful train ride across the countryside, the North Yorkshire Moors Railway, a picturesque journey on a train pulled by steam! No visitor should visit Pickering, or indeed Yorkshire, without enjoying a day out on this excellent little railroad, and spending a little time either before or after the journey looking at the engines or wandering around the bookshop and information centre.

PICKERING lies 25 miles north of York on the A170.

PORT SUNLIGHT
Lancashire

Those who think that 'New Towns' are a post-World War II idea, should take a look at Port Sunlight, a garden village on the banks of the River Mersey, built to house the employees

The lush greenery around Richmond, seen from the hilltop castle

of the local Unilever soap factory. This is no collection of grim tenements, but an architectural storehouse of styles. Nearly all the buildings are now under Conservation Orders and the Lady Lever Gallery in the centre contains a unique collection of works by the Pre-Raphaelites, and Wedgwood, as well as ceramics, tapestries, furniture and needlework. The over-riding impression though, is of a pleasant place to live, after the day's work is done.

PORT SUNLIGHT lies south-west of Manchester on the banks of the River Mersey.

RICHMOND
North Yorkshire

Those who journey through England hoping to find some splendid little-known places, will chance upon Richmond with delight. This is probably the most attractive town in Yorkshire, enhanced by the great ruins of the castle which stand on the hilltop above the town. It was built in 1071 and legend has it that King Arthur sleeps somewhere beneath it. Certainly the views from the ramparts are fabulous, looking out over the town below and Swaledale.

Richmond is seamed with little wandering alleyways, called *wynds*, and has the largest market square in Britain, attractive walks by the River Swale, and a great abundance of medieval architecture. Do not leave without visiting the museum of the local regiment, the Green Howards, or the Grey Friars Church.

RICHMOND lies 3 miles north of Catterick.

RIPLEY CASTLE
North Yorkshire

Set in a deer park, with lakes and walled gardens, is this ideal kind of English manor house, built on medieval foundations and still retaining part of the old walls before the more modern, Tudor and Jacobean additions.

The castle has been owned by the Ingelby family for nearly 700 years and is now open to the public regularly throughout the summer. The library is beautiful and the Knight's Chamber contains a fine collection of armour worn by the family during the Civil War. This chamber also contains a Priest's Hole built into the fabric to hide fleeing Jesuits in 1544 and only discovered in 1964. There are family portraits, fine furniture, beautiful panelling, and all one could wish to see in an historic house that is also a family home.

RIPLEY CASTLE lies 7 miles north-east of Harrogate on the A61.

RIPON
North Yorkshire

Ripon is a rather stark city, less cluttered than many of the older English towns. Old it certainly is, for St Wilfrid founded the great cathedral here in about AD 670, and parts of that building can be seen under the crypt of the 12th-century building. Ripon Cathedral is particularly interesting because it contains all the main periods of Gothic architecture on a Saxon foundation, and it all blends together most successfully.

Apart from the cathedral there are old houses round the market square, where the Town Crier or Wakeman can be heard every evening, sounding his horn. There are also pleasant walks along the banks of the River Skell.

RIPON lies 26 miles north of Leeds on the A61.

SCARBOROUGH
North Yorkshire

Scarborough is one of those towns with some claim to the title of the 'Queen of English seaside resorts'. The town lies around the wide bay and is protected by a Norman castle.

Donkey rides are still popular on Scarborough's sands

Its present role as a seaside resort and spa town dates back to the 1620s when mineral springs were discovered, but the town really developed in the last century when many of the great hotels were built on the cliffs and behind the harbour.

The castle, which catches every eye, was built by Henry II, but it was besieged regularly thereafter, constantly damaged and as constantly rebuilt. Down in the town, St Mary's Church is medieval, but St Martin's-on-the-Hill is last century Gothic, a shrine to the Pre-Raphaelites, with work by William Morris, Burne-Jones, Rossetti and Ford Madox Brown.

SCARBOROUGH lies 20 miles east of Pickering on the A170.

THE SETTLE TO CARLISLE RAILWAY
Yorkshire

The Settle to Carlisle Railway is a marvellous train ride, taking the traveller across some beautifully scenic parts of Yorkshire, into the Pennines and out into Cumbria. The ride actually begins in Leeds, from where the line goes north-west to Keighley and Skipton before the true trip begins at Settle. From here the track continues north and west across the wild moor, through Blea Moor Tunnel which is 2649 yards long, over Dent Head Viaduct to Dent, then through Garsdale and out into Cumbria.

All along the way the views are quite superb.

SETTLE lies 12 miles north of Skipton on the A65.

SKIPTON
North Yorkshire

Set in the heart of the Yorkshire Dales, Skipton is one of those perfect

 Skipton Canal basin is filled with a variety of craft

little country towns, with all the traditional elements intact, a castle, an old church, a weekly market. The town centres on a wide main street, down which sheep were driven in years gone by. Nearby, in Chapel Street, is the 13th-century High Corn Mill, now a museum, well restored and still producing working grindstones and waterwheels.

Skipton Castle, near the parish Church, is protected on one side by the sheer rock wall. The bulk of the castle dates from 1310-15, when it was built by the Clifford family, who lived here until the 1950s. In the Civil War they held this castle for King Charles, but after his execution the walls and towers were blown up — or 'slighted' — to make the castle useless as a stronghold.

SKIPTON lies 26 miles northwest of Leeds.

SLEDMERE HOUSE
North Yorkshire

Sledmere is the Yorkshire home of the Sykes family and a magnificent example of the Georgian country house. It was built in 1751 and enlarged in 1758, gutted by fire in 1911 but rebuilt to the original designs after World War I. It is still lived in by the Sykes family, and much of it is open to visitors, which is a good thing as there is a lot to see including Chippendale and Sheraton furniture, paintings, porcelain and plenty of statues as well as a finely tiled Turkish Bath.

SLEDMERE HOUSE lies at Driffield, 16 miles from Bridlington.

SPEKE HALL
Lancashire

Speke is better known today as the location of Liverpool's local airport, but those wishing for a flight would do well to leave a little earlier and spend an hour or so looking over the attractions of Speke Hall. Speke is a Tudor mansion and has all the attractive Tudor features, with lots of half-timbering, four wings surrounding a central courtyard, a massive Great Hall, and some excellent panelling and carving. The courtyard contains two very ancient yew trees, while the interior has an unusual collection of Victorian household artifacts, William Morris wallpaper and some fine tapestries.

SPEKE HALL lies in Merseyside, 8 miles south of Liverpool on the A561.

WENSLEYDALE
North Yorkshire

Wensleydale in the Yorkshire Dales National Park is among the most enchanting of all the Yorkshire dales and the one to visit if time is pressing, for it contains some beautiful villages and the great natural attraction of that famous waterfall, Aysgarth Force.

Wensleydale is a green and graceful place, and the calm valley of the River Use, which flows through here into the Yorkshire Ouse. Just to the south at the eastern end lies Middleham Castle. Leyburn is the only town of any size and there are some beautiful villages. Wensley itself has a fine church with some excellent brasses, while nearby Bolton Castle was built by the Scropes in the reign of Richard II.

Aysgarth is very central, especially for walkers, and two essential sights here are the Force and the church which stands nearby. This done, climb the sides of the dale, up onto one of the small rocky escarpments and just enjoy the views which are among the best in Yorkshire.

WENSLEY lies 20 miles north of Ripon on the A6108.

The Shambles, one of York's picturesque shopping streets

WHITBY
North Yorkshire

This is the place where Captain Cook started his sailing career and some items of Cook memorabilia can be seen at the Whitby Literary and Philosophical Society Museum in Pannett Park. Whitby is a lovely little fishing port on the River Esk and there was a town here in Saxon days. Whitby Abbey was founded by the Saxons long before the Norman Conquest.

Today, Whitby is still a fishing port and a popular centre for visitors who use Whitby as a touring base.

WHITBY is on the coast, 45 miles north east of York.

YORK
North Yorkshire

If the North of England can be said to have a capital, York is the city which would command that title. Indeed, York was, and still is, the second city of the Island Kingdom. In many ways it is more interesting, historically, than London, for it has retained more of the feel of a capital city, and contains, even today, a host of historical artifacts; Jorvik, for example, recalls the capital of the Danelaw, when the Vikings ruled at least half of England. York was an outpost of Denmark until 944, and the very street names recall that connection — the Shambles, Walingate, Micklegate, Whip-ma-Whop-ma Gate, where thieves were flogged — no one could visit York and doubt that this was once a very important city of the Kingdom, as indeed it still is today.

York is not a place to rush through in a day or two; it has a host of places to slow the traveller down, from the famous Minster to the modern Railway Museum, which opened in 1975 and is the finest in Western Europe. Half a day at least must be allowed for this museum alone, and then there are the marvellous parish churches, the shops and markets, the medieval Merchant Adventurers' Hall in Fossgate which dates from 1360, and as an introduction to the County of Yorkshire, the Yorkshire Museum in Museum Street. A full weekend, or better still, a week, is the minimum time necessary for a real visit to York.

The Cathedral Church of St Peter is one of the largest English cathedrals. The present Minster has now been restored after years of work, and a sudden fire in 1984. The present building dates from the 1250s, but it stands on the site of a much earlier church built by Bishop Paulinus about AD 625.

York Minster is simply magnificent, a poem in stone. It celebrates the glory of Gothic architecture, and in the West window, the end of the Wars of the Roses; examine the glass closely, for those alternating emblems of that long Plantagenet conflict can be seen. Earlier work includes the 13th-century Five Sisters window, the Chapter House of the monks, and a marvellous medieval vault.

Restoration work carried out in the 1970s and 1980s has ensured the survival of York Minster for centuries to come, and revealed the Roman roots of the ancient City of York itself.

YORK MINSTER dominates the City of York, which lies 26 miles south of Leeds.

Berwick on Tweed
Lindisfarne
The Farne Islands
Bamburgh
Chillingham Castle
A1
Craster
Alnwick
Warkworth Castle
A 68
NORTHUMBERLAND
A1
A 68
Hadrian's Wall
A 69
Lanercost Priory
TYNE & WEAR
Washington
Talkin Tarn
Alston
A 595
M6
A6
A 68
Durham
A 19
CUMBRIA
COUNTY DURHAM
Keswick
Derwentwater
Appleby-in-Westmorland
Ullswater
Raby Castle
A 66
Barnard Castle
A 1(M)
Buttermere
St Bees
Wastwater
Grasmere
Hardknott Pass
Ravenglass
Windermere
Coniston
Bowness
Lake Windermere
Garsdale
M6
0 10 20 30 km
0 10 20 miles

The Northern Counties

An evening view across Derwentwater from Friars Crag in the Lake District

THESE SPLENDID NORTHERN counties lie around the Scottish March and between them straddle the Pennines. Northumberland and Cumbria are the northern buttresses of England, and the traveller who wanders about here, along those narrow, high-banked roads, will soon find plenty of evidence of a not far distant warlike past.

Across the neck of land between Newcastle and Bowness lies Hadrian's Wall, that great 1st century fortification built to keep the warlike Picts in their place. The old kingdom of Northumbria, which now, as a tourist zone includes Northumberland, Durham and Tyne and Wear, is studded with fine castles, like mighty Bamburgh or splendid Alnwick, where the Percys have lived for over 700 years.

A little to the north and a 'must' on any traveller's itinerary, lies Lindisfarne, the Holy Island, which is in fact only a part-time island for a long curving causeway links this historic spot (where St Cuthbert and his monks brought Christianity to the pagans), to the mainland a mile away, while further out to sea the Farne Islands lie flat against the ocean, loud with the shrieks and cries of seabirds.

Across the Pennines, between the Eden Valley and the coast of the Irish Sea lies the famous Lake District of England, a beautiful spot with hills and lakes and little towns, a mecca for climbers and walkers, a place of poets and a popular haunt of holidaymakers.

Alnwick Castle, border fortress and ancestral home of the Percys still guards the road south from Scotland

ALNWICK
Northumberland

Alnwick (pronounced Annick) is one of the greater border towns of England, built during centuries of war and the home of the Percy family, who have lived and ruled in Northumberland for over 600 years, Harry Hotspur being one of the most famous of that warlike line and the one immortalised by Shakespeare in *Henry IV*.

Alnwick is still a fine medieval town, full of narrow streets with antique names, Bondgate, Narrowgate, Bailiffgate and so on. There is a curious pub known locally as the 'Dirty Bottles', which is said to be haunted, and many interesting tombs in the Church of St Michael. Looking about the town it is interesting to note how many buildings bear the stiff-tailed lion emblem of the Percy's. Only one of the town gates is still standing but this, luckily, is the 14th-century Hotspur Tower, and the past is still fairly intact in the streets around the market.

Unlike all too many English castles, Alnwick is still inhabited, and by the family who built it, the Percys. The Percys first came to Alnwick Castle in 1309, and their first task was to strengthen it against the Scots. The process of building and alteration has never stopped and over the centuries Alnwick has been transformed from a breezy medieval fortress into a great country home. Robert Adam restored it, but most of it is still outwardly a 14th-century castle, topped by a series of figures along the battlements, which look like living people when seen from far off across the water meadows down by the River Aln, which was the original idea. In the last century the interior of the castle was converted into the style of the Italian Renaissance, which is still quite pleasing and very effective. Much of the castle is open to visitors in summer, when the guardroom, keep and library can all be visited. The walls are furnished with works by the Italian masters, Titian, Tintoretto and so on; the family coach rests in the stables and a museum of the Royal Northumberland Fusiliers is well worth seeing. All in all, a place which is full of history and interest for the visitor.

ALNWICK is 30 miles south east of Berwick-upon Tweed.

ALSTON
Cumbria

Alston is a quaint, sturdy little town, the highest market town in England. Built of grey stone, it's a place that seems to topple about in all directions, with no street staying on the level for long. Alston has no major sight to compel the visitor's attention; the attraction is all in the setting and in seeing a town which retains a strong independent and very English air. The railway station is worth inspection, as are the old churches, and the roofs of the solid stone houses, but that apart, Alston is a good touring centre, west to Weardale and Durham, north to the Border across the moor, or south towards Melmerby, where at Melmerby Fell there is a famous view across the vast valley to the west, even to the distant fells of the Lake District.

ALSTON lies 20 miles south-east of Brampton by the A689.

APPLEBY-IN-WESTMORLAND
Cumbria

Before the English counties were regrouped in the 1970s, an act that many people still regret, Appleby was the capital of the historic and beautiful county of Westmorland. It lies on the River Eden, a sparkling stream, and has long been the centre for a famous Horse Fair held on the second Tuesday and Wednesday in June. It is dominated by Appleby Castle, still a private home, where the grounds contain a centre for the Conservation of Rare Breeds. This castle and the town which runs down the hill from the gatehouse belonged to the Clifford family, and Lady Anne Clifford restored both to their present state after Cromwell sacked the place during the Civil War. At the foot of the High Street, which is lined with old inns and houses, stands the parish church, a splendid example of a medieval Gothic building. Little Appleby offers good walks along the River Eden and into the Western Pennines and is, taken all together, a most attractive example of an English market town.

The Moot Hall of Appleby-in-Westmorland, seen from the castle gates

APPLEBY-IN-WESTMORLAND lies on the A66, 21 miles south-east of Penrith.

BAMBURGH
Northumberland

Bamburgh is dominated by the great formidable pile of Bamburgh Castle, one of the mightiest castles in England, which in one form or another has guarded this shore and held the border against the Scots these thousand years or more. From the ramparts visitors can see out to the Farne Islands, now a famous bird sanctuary, to Lindisfarne, inland to the Cheviot Hills, north to Berwick. Bamburgh was once the capital of the Kingdom of Northumbria, in the days of the Saxons, and, later, a great fief for the Percy's, that noble family which still lives in Alnwick. The castle looks at its best when seen from the shore, but is impressive inside, with narrow passages, crenellated walls and a magnificent medieval King's Hall. The castle is open to visitors in summer.

The village at its foot, all that remains of the old town, is the birthplace of the heroine Grace Darling, who rowed out one stormy night to rescue the sailors from the ship wrecked on the Longship's Rock. Grace is buried in the churchyard, and the village contains a small museum dedicated to her exploits and the lifeboat service.

BAMBURGH lies on the B1342 30 miles south of Berwick-on-Tweed.

BARNARD CASTLE
County Durham

Barnard Castle is a rather lush little town on the River Tees, overlooked, as the name implies, by the now ruined walls of a medieval castle; what town around here could have survived without one? A Norman, Guy de Baliol, began the castle shortly after the Conquest, and his nephew, Barnard, gave his name to the town which grew up around it; his son John founded Baliol College, Oxford, while *his* son, John, became King of the Scots in 1292. Much later, Barnard Castle, town and fortress, came into the possession of Richard III. Of the fortress which he must have visited, only the keep and part of the walls remain, but the town is full of interest, with a sweeping 15th-century bridge and a great deal of attractive 17th and 18th-century architecture. The Church of St Mary, situated in the Market Place, is basically Norman though much restored by the Victorians. The Bowes Museum, a little to the west of Barnard Castle is really superb.

BARNARD CASTLE lies 6 miles south-west of Raby Castle.

BERWICK ON TWEED
Northumberland

Berwick is a fortress city, built and rebuilt against the Scots. The last rebuilding, by Queen Elizabeth I, was practically unnecessary, as her successor, James I of England and VI of Scotland, united the two kingdoms with the result that the Elizabethan walls of Berwick are still intact.

Today, Berwick is an attractive little town, straddling the Tweed and overlooking the small harbour which is still full of fishing boats.

The sparkling Tweed is spanned by three bridges and the town, not surprisingly, is a centre for salmon fising. Lovers of fortifications will enjoy walking along the walls, or visiting the old barracks, designed by John Vanburgh, the architect of Blenheim Palace, which is the oldest example of military barracks in Britain. In previous centuries, and for a long time afterwards, troops were simply billeted on the local population. The museum of the local regiment, the King's Own Scottish Borderers, is also well worth a look for those who enjoy militaria.

BERWICK is on the A1 road, 46 miles north of Alnwick.

BUTTERMERE
Cumbria

Buttermere is one of the most beautiful, wildest and least touristic lakes in Lakeland, a place which draws the visitor again and again, with the ceaseless attraction of its views, which lie vast and varied on every side. If one must choose, the view across the lake to Haystacks from Buttermere village will take some beating. The Buttermere fells rise up to the north-east, Red Pike and High Stile are open to the south and west, making this lake — or mere — the perfect starting point for walkers who enjoy trekking in the high fells.

By that spit of land which is all that prevents Buttermere joining up with Crummock Water, lies the tiny village of Buttermere. Another nearby attraction is Scale Force waterfall, which curves down for 120 ft between the rocks, the highest waterfall in the Lake District.

Buttermere is peaceful and beautiful at any time of the year, but perhaps most of all in autumn, when the fall colours tint the leaves and the bracken on the fells. Those who don't want to walk on the fells or around the lake can view all this from the B5289 road which runs along the eastern shore and is well supplied with parking places.

BUTTERMERE lies on the B5289, 20 miles south-west of Keswick.

The small church, or chapel, in Buttermere village

CHILLINGHAM CASTLE
Northumberland

By now it must be obvious that Northumberland is a land of castles. This county is full of them, topping every hillside, guarding fords and road junctions, each a relic of the long centuries of Border strife. At Chillingham Castle however, the attraction is even more interesting and yet still alive. The white cattle of Chillingham have been there since prehistoric times. The present animals are directly descended from a herd of wild white oxen which once roamed the great forest which covered Northumberland in the past. The present herd is quite small, only 40 beasts, but some trick of genetics evades inbreeding and under the protection of the Chillingham Wild Cattle Association, they continue to thrive. The cattle really are wild and can be dangerous to approach on foot, but they remain a rare survival from a far-off age, and long may they continue to live here.

CHILLINGHAM CASTLE lies 2 miles east of Wooler off the A697.

CONISTON
Cumbria

Coniston draws visitors from all over the world. Those who like the high fells will be attracted by the loom of that famous peak, The Old Man of Coniston, which any able-bodied walker can reach by following the footpath signs from the village centre. Two miles to the north-east lies the gem-like little lake of Tarn Hows, one of the most photographed lakes of Lakeland. Coniston Water itself lies only a short walk from the village.

Coniston Water is just over five miles long and perhaps best known in

Low cloud can mark the top of Coniston Old Man, even on a summer's day, and snow often lingers on the summit

recent years as a centre for various attempts on the world waterspeed record. Sir Malcolm Campbell set up a record here in 1939 and his son Donald was killed here during another attempt in 1967. Although other boats using the lake are restricted to 10 mph, it is still used for waterspeed record attempts when the speed rule is temporarily waived.

Such events apart, Coniston is also noted as the home of the Victorian writer John Ruskin, who lived at Brantwood on the eastern shore from 1871 to 1900. His house is much as he left it and open to the public. Ruskin was buried in Coniston village and the Ruskin Museum in the village centre contains many other Ruskin memorabilia, including the pall which covered his body.

It is possible to cruise on the lake, notably on the steam yacht *Gondola*, built in 1859 and beautifully restored by the National Trust.

CONISTON lies 5 miles from Hawkshead on the B5285.

CRASTER
Northumberland

Craster is a fishing port, famous for well-smoked kippers. In itself it is only a little place, a couple of streets full of fishermen's cottages, a small harbour, and the rich tang of the sea, but the coastline here is very beautiful.

A mile to the north, a half-hour stroll across the springy turf, lies the great gaunt pile of Dunstanburgh Castle, its foundations resting on the wave-washed rocks, a marvellous sight in the early morning as it looms up out of the sea mist. Needless to say, it is haunted, by Sir Guy who roams the Great Hall seeking his lost lady. Dunstanburgh is the largest of all the Border castles, and was built in 1313 by the Earl of Lancaster, later passing to John of Gaunt. During the Wars of the Roses it was under constant siege, and by the mid-16th century it was as we see it today, in ruins. It remains a wonderful sight; Turner painted it several times and no one should pass this way without seeing it.

CRASTER lies 6 miles east of Alnwick.

DERWENTWATER
Cumbria

People will argue about which is the finest of the lakes of the Lake District, but all would concede that Derwentwater takes some beating. The name has a Scandinavian root and is said to mean, 'the lake of the river which is fringed by oak trees', as indeed it is. Derwentwater is the widest lake, one-and-a-half-miles across, three miles long and 72 ft deep at the deepest point. Elsewhere it shallows to less than 18 ft, so it freezes during most winters and therefore attracts skaters by the score. There are five islands on the lake plus one odd one known as Floating Island, which is composed of weeds and sunken vegetation and only pops up now and again, propelled to the surface by marsh gas.

The lake can be viewed from many parts around the shore and is encircled by various roads but to enjoy it, as well as see it, take a cruise on one of the launches which ply from Keswick. It is possible to get off at one of the landing stages around the shore, go for a walk and pick up another boat later on or from another stage — a good way to pass the day.

DERWENTWATER runs south from Keswick to the Jaws of Borrowdale.

DURHAM
County Durham

Durham is a splendid town, set above the River Wear, one of the undebatable glories of the North. It is an old town, founded by the monks who fled from Lindisfarne, and by the early years of the 11th century there was a castle and cathedral here, protecting the shrine of St Cuthbert. The present city, although still small and unlikely to expand because of its siting, is one of the most delightful little cities in England, crowned by the great soaring Romanesque cathedral. The river cutting past the cliffs is littered with punts and rowing boats in summer; the narrow streets are filled with visitors from all over the world, and students from the University give the town a youthful, lively air.

Just off Palace Green stands Durham Castle, built by William the Conqueror to nail down the north. This was the residence of those formidable clerics, the Prince-Bishops of Durham, who founded the University. The castle is full of interesting relics, most notably the 13th-century Great Hall, hung with banners and lined with armour. The Keep was refashioned in the last century and is now a dormitory for students, but much of this great castle is as it was during the Middle Ages and, with the cathedral, it provides a fine central core to this lovely old and interesting town.

The present Durham Cathedral stands on the site of the former Saxon cathedral, and being built by the Normans in 40 years from 1093, is basically Romanesque, or as the English call it, Norman, although there are many overtones of the later Gothic style.

The site is spectacular and the best view of the cathedral is the one which displays this to the full, the view from the opposite bank of the River Wear to the great building high up on the cliff. Inside the effect is still overwhelming, with a soaring vault above the thin ribbed columns, and a number of later, elegant chapels, notably the 12th-century Galilee Chapel. The Nine Altars Chapel was added a century later and the Central Tower, which gives tremendous views over the town below and the countryside round about, was rebuilt at the end of the 15th century. Three essential and rare attractions at Durham Cathedral are the Sanctuary Knocker on the North Door which, if he could grasp it, gave protection to the fleeing criminal (sanctuary rights were abolished in 1540); the painted wooden coffin of St Cuthbert, carried here by the monks of Lindisfarne nearly a thousand years ago; and in a chapel off the nave, a bullet-riddled cross which once topped the Butte de Warlicourt on the Battlefield of the Somme, a height captured by the Durham Light Infantry in the Autumn of 1916.

DURHAM lies 20 miles south-east of Sunderland, just off the A1 (M) motorway.

THE FARNE ISLANDS
Northumberland

The Farne Islands are easier to see than to reach, but those who are able to make the journey across the tumbled waters of the North Sea will not regret the trip, for these islands which now belong to the National Trust are one of the country's most important seabird sanctuaries, a migrant stop in the spring and autumn and a place full of breeding colonies for gulls and terns. Seen from Lindisfarne or Bam-

Dove Cottage is now part of the Wordsworth Museum

burgh they look like nothing more than a litter of flat, grey rocks on the silver face of the sea, but when the easterly gales sweep down from Norway, the Farnes become a fearsome spot and many ships have come to grief on the Longships Reef.

Visiting is not permitted during the breeding season in the late spring, but otherwise there are boat trips out to the Farnes from the little port of Seahouses. Getting ashore on the Farnes is a tricky business and depends on the tides and sea-state, but the trip there is still well worth while.

SEAHOUSES lies 3 miles south of Bamburgh on the B1340.

GRASMERE
Cumbria

No visitor to the Lake District gets away without visiting Grasmere or, to be more exact, Dove Cottage, where Wordsworth lived from 1799 until 1808, during which time he wrote most of his more famous poems. Wordsworth loved Grasmere and the Lake District and made it famous, but people would go there anyway for it is a beautiful spot. The mere at Grasmere lies in the valley and reflects the shape and colours of the surrounding woods and fells. A main road divides the village from Dove Cottage, and while Wordsworth's home and the newly opened Grasmere & Wordsworth Museum have to be seen, the village itself is well worth wandering around, especially the Church of St Oswald which dates from the 13th century.

GRASMERE lies 3 miles north of Ambleside on the A591.

A view over Grasmere in the early spring, when snow still lies on the top of the nearby fells

Lakeland sports

Those who seek visible evidence that the Lake District of Cumbria is a place apart need do no more than visit a Lake District Summer Show. These are held throughout the summer in many different parts of the Lakes and have all the usual trappings of an English agricultural show, plus some unique local events.

There are stands and sideshows, pony club events, horse jumping and parades of cattle and livestock, but while all this is going on, the eye is caught and the attention held by the local people competing in some distinctly local attractions; fell-running and Cumberland and Westmorland wrestling.

Fell-running requires the stamina of a marathon runner with the nerve and ability of a mountain goat, for the competition, if simple, is never less than demanding; just spring up some 2000 ft or so to the top of the nearest fell — or maybe two — and then run back to the finish, over rocks and scree, along steep sheep tracks with the risk of a broken ankle at every stride. This is a sport you must be born to, and the winner is usually a local lad.

The wrestling is said to date back to Viking times and is now an exclusively Cumbrian affair which looks like a simple trial of strength but actually calls for considerable skill. The wrestlers face each other in a small arena, grasp their hands behind each other's backs and try to topple their rival over, while breaking his hold, and the one who comes out on top is the winner. It may sound simple, but it is not for the feeble and no Lakeland Show can be without a contest between the local strong men.

Local men test their strength in a bout of Westmorland wrestling

HADRIAN'S WALL
Northumberland

The great Roman Wall, built against the Picts who then lived in Scotland, by the Roman Emperor Hadrian in the decade from AD 120, runs for 72 miles from Bowness-on-Solway in the West, right across the narrow neck of Northern land to Wallsend on the River Tyne. It is undoubtedly one of the finest artifacts in England, and one of the most splendid examples of the energy of the Roman Empire still standing in Western Europe. The Wall is in an excellent state of preservation and recent excavations have revealed more of the way-stations and garrison camps which once supported the Wall itself. When completed the Wall was 15 ft high, 10 ft wide, and protected by a steep ditch.

It was built to mark and guard the northern-most boundary of the Roman Empire, and was garrisoned for nearly 300 years until the legions were withdrawn in the early years of the 5th century AD. The Wall can be viewed, running along the crest to the North, at many points along the B6318 west of Newcastle, and one particular excavation which has a museum, is the remains of the Roman fort at Housesteads.

Housesteads stands to guard one of the more vulnerable places, and since the foundations are intact, the visitor can see just how large and strong these Roman forts could be.

HADRIAN'S WALL lies west of Newcastle-on-Tyne, close to the B6318.

HARDKNOTT PASS
Cumbria

The Hardknott Pass is one of the most dramatic parts of Cumbria. It runs east-west along and astride the

Hadrian's Wall stretches along what was once the northernmost boundary of the Roman Empire

narrow, intensely winding road which leads from the western coast at Ravenglass, beginning at Eskdale and climbing over and down to little Langdale through its eastern extension, the almost equally famous and spectacular Wrynose Pass. This road, small as it is, is very ancient, following the path of a prehistoric track and a later Roman road. On the top of the pass, at 1290 ft, stands the ruin of a Roman fort, but the real attractions are scenic rather than historic. This road can become very crowded indeed in summer, and should only be attempted by motorists with strong nerves and good brakes, especially if heading towards the west.

THE HARDKNOTT PASS lies east of Eskdale, above Boot village.

KESWICK
Cumbria

Overlooked by one of the magnificent peaks of the region, Skiddaw, Keswick is the 'capital' of the Northern Lakes and therefore ideally situated at the northern end of Derwentwater and the foot of Skiddaw. Once a mining town and a market for the people who lived and farmed on the local fells, it is now devoted to tourism. Attractions in the town include the Keswick Railway Museum and the church-like Moot Hall which stands in the very centre of the town. Sights round about include the prehistoric Castlerigg Stone Circle, dating from circa 1400 BC, which lies to the east, and the Friars Crag Nature Walk. That apart, no one visits Keswick without taking the long walk up towards the peak of Skiddaw where, according to the poem, the beacon flared to "......wake the burghers of Carlisle" when the Spanish Armada was sighted in 1588. Skiddaw is topped by snow even in summer, and if the day is clear the views from the top are quite outstanding.

KESWICK lies 27 miles west of Penrith on the A66.

LANERCOST PRIORY
Cumbria

Not much remains of the once vast priory of Lanercost, but what there is should provide an interesting visit, and the setting itself is beautiful. It lies some five miles north-east of Brampton, and was founded in 1169 for the Augustinian priors by a local lord, Robert de Valibus. Most of the priory was destroyed in the Reformation, when the community was dispersed, but the priory church, in the local dull red sandstone, was retained to serve as the parish church, which it still does, attracting a considerable congregation every Sunday and a host of visitors throughout the summer months. In the ruins of the priory, behind the church, note the 13th-century cross in the Celtic fashion, as well as the much later Tudor bridge which spans the river nearby. The Roman Wall passes close by and can be seen at Banks.

LANERCOST lies 5 miles north of Brampton, on a minor road off the A69.

LINDISFARNE
Northumberland

Lindisfarne, the Holy Island, lies across a causeway from the coast of Northumberland, and is cut off by the tide twice a day. The original monkish settlement here was made by Celtic monks from Iona, led by St Cuthbert, but the Vikings drove them out eventually, and the present priory, though in ruins, dates from the 11th century and was built by the Benedictines.

Lindisfarne is a magic spot, a paradise for birdwatchers, historians and all lovers of the rare quiet places, though it can get very crowded indeed in high summer. The church was once part of the priory, and contains replicas of the Lindisfarne Gospels, as well as many memorials and stained glass. From the lower end of the churchyard, Lindisfarne Castle stands framed in a raised window arch, high on a spur of dolerite rock, a wonderful sight.

The castle, which is now in the hands of the National Trust, was built in the 16th century, mostly from stone filched from the abbey after the monks were dispersed by the Reformation. In its time it served as a border fortress, barracks and coastguard station.

At the turn of the century, the ruined castle was restored by the architect Edwin Lutyens, and the result is this splendid castle of the present day. From the upper battery there are vast views to Bamburgh, Berwick or the Farne Islands, far out to sea, while just across the valley lies the unique walled garden created by Lutyens' collaborator, Gertrude Jekyll. From up there, with birds circling overhead and seals cavorting in the waves offshore, Lindisfarne, the Holy Island, is a marvellous sight.

LINDISFARNE lies offshore, 16 miles south of Berwick.

Lindisfarne Castle, high on a spur of rock

RABY CASTLE
County Durham

Raby Castle is a famous pile and still topped by nine crenellated towers. It is regarded as one of the largest and best preserved castles in the North of of England. There was a fortification here in the time of King Canute, but the present castle was begun by the Nevilles in the 12th century and remained in their hands until 1569 when the last Neville was executed for attempting to put Mary, Queen of Scots, on the throne of Elizabeth I. The Great Hall of Raby Castle was built to seat 700 knights. In 1626 Raby was sold to Sir Henry Vane and his descendants live here to this day, having carried out alterations from time to time. In the 18th and 19th centuries it became a country seat, and therefore displays a range of periods in both architecture and decoration. The Neville Gate and the towers are 13th century, the gardens are 18th and 19th, with trim yew trees and a spread of oaks, while the interior is Victorian. The castle is open to visitors on most days during summer.

RABY CASTLE lies 1 mile north of Staindrop, and 9 miles south of Bishop Auckland on the A688.

RAVENGLASS
Cumbria

Ravenglass, which is now reduced to little more than a single street of old houses with one good hostelry, The Pennington Arms, was once a famous place and a major port. When the Romans came to subdue this part of Britain, Ravenglass was their port, and it remained important until the middle of the last century, served by the Ravenglass and Eskdale Railway which brought ore down to the quay here and is still running tourist trains

Ullswater, most attractive of the lakes, winds its way north from Glenridding to Pooley Bridge

into the fells. Sights in the town include the harbour, the famous 'gullery' where seabirds nest in spring, Muncaster Castle just up the hill, and the Railway Museum in the station. The High Street is full of fine old houses and little shops.

RAVENGLASS lies on the west coast of Cumbria, 23 miles south of Whitehaven.

ST BEES
Cumbria

St Bees is a straggling little coastal village of whitewashed houses, set beneath the jutting promontory of St Bees Head, one of the great viewpoints on the West Cumbrian coast. The town is chiefly noted for the public school there, but there is also a fine beach and a host of seabirds nest on the high cliffs of St Bees Head. A little to the north lies Whitehaven, once a coal and iron-ore centre, and to the east lies the old town of Egremont.

ST BEES lies 4 miles south of Whitehaven on the B5345.

TALKIN TARN
Cumbria

Talkin Tarn is a most attractive little lake, set a mile or so south of the small market town of Brampton. As is often the case in Cumbria, the setting is the reason for going there, and the setting for Talkin Tarn is quite superb. There are good walks around the shore, a delightful hotel for lunch or dinner, even a golf course, but most people will be quite content to sit under the trees and just gaze out across the waters of the lake.

Brampton, a mile to the north, is a pleasant, unassuming town, a popular place with ramblers and cycle-tourers, convenient for those who wish to explore the western end of Hadrian's Wall.

TALKIN TARN lies a mile south of Brampton on the B6413.

ULLSWATER
Cumbria

Ullswater is the most varied of the lakes. Others may be a little larger or more dramatic, but none has as much to offer the visitor. Ullswater is a winding lake, over seven-and-a-half miles long, and nearly a mile wide at its broadest point, running north from Glenridding and Patterdale to Pooley Bridge. It can be viewed and enjoyed from the A592 road which runs along the western shore, or better still, from the deck of one of the many boats which cruise about the water from Pooley Bridge down to Glenridding and back again throughout the summer. A good place to start any visit to Ullswater is in the straggling little village of Glenridding, reached from the south over the Kirkstone Pass, from where the road falls past the little lake at Brotherswater down to the big lake by Patterdale.

ULLSWATER lies 20 miles north of Windermere on the A592.

WARKWORTH CASTLE
Northumberland

Warkworth is another very splendid example of a Border fortress. It was built originally by Henry, son of King David I of Scotland, and that should not be too surprising for many Scots kings held land in England. Henry was made Earl of Northumberland in 1139, but dispossessed in 1157 when Henry II gave the castle to Roger de Stuteville, whose family built much of the present castle before it passed to the Percys in the 14th century.

Much still remains. The Grey Mare's Tail is a fine tower, pierced by cross-bow loopholes, the Lion's Tower still guards the entrance, and the Keep has been transformed into a house, although it is still reminiscent of the Middle Ages. A tour of the castle can take in the guardroom, the pantry, the Great Hall which Shakespeare featured as the home of Harry Hotspur in *Henry IV: Part 1*; visitors can also see the castle chapel which looks out across the tumbled waters of the North Sea towards Coquet Island.

WARKWORTH lies 10 miles south of Alnwick on the A1068.

WASHINGTON
Tyne & Wear

Washington is little known outside the district, and deserves to be more famous than it is, for Washington Old Hall, a Jacobean residence which lies just on the outskirts of the village, was the ancestral home of George Washington, first President of the United States.

The Washingtons arrived here in 1183, although they lost the manor of Washington in 1376 and dispersed throughout the kingdom, which accounts for the fact that various other parts of Britain lay claim to the patriotic founder of the U.S.A. The house, which is open to the public during the summer months, contains a fine collection of 17th-century furniture and much Washington memorabilia.

WASHINGTON lies 5 miles west of Sunderland on the A1231.

WASTWATER AND WASTWATER SCREES
Cumbria

Few places in the Lake District are as fine and memorable as Wastwater, and the Lake District is full of memorable places which remain fixed in the mind's eye. To see this long, dramatic lake at its best, go to the south-western end. From there the lake and the steep, dark screes which plummet down to the water along the south-eastern shore are quite overwhelming, while through the gap in the hills to the north one can see the central fells and the distant peak of Great Gable.

Wastwater is three miles long and at 258 ft, the deepest lake in England. The Screes peak at the 2000 ft mark, and although they look practically sheer at a distance, they lie back quite a bit if you cross the lake and take the footpath which runs along the south-eastern shore. At the head of the valley (a favourite spot for hill-walkers) lies The Wasdale Head Inn, where the Will Ritson Bar commemorates a man who claimed to be the biggest liar in England. Wasdale, the valley which contains Wastwater, is a beautiful spot, but tends to be empty of people even in the height of summer.

WASDALE lies 15 miles north-east of Ravenglass, off the A595.

WINDERMERE
Cumbria

To say Windermere 'Lake' is tautological, for a mere *is* a lake, but we refer here to the lake (or mere), not the town on its banks. Windermere is England's largest lake; ten and a half miles long, a mile wide, over 200 ft deep. These waters are cold and clear and contain a rare fish, the *char*, a popular local delicacy, and it is the lake which draws most visitors to this popular spot in the Lake District. They come to swim, paddle, sail, row, windsurf or cruise, and in summer time certainly, the surface of Windermere must creak under the weight of passing craft.

Steamers run up and down the lake from Bowness to Ambleside and south to Lakeside, but other visitors stay closer to the centre around Bowness and simply take the guided tour across to Belle Isle, a 38 acre island just offshore with the only truly round house in England on it. A ferry cuts across the lake, taking visitors to the opposite bank for a visit to Beatrix Potter's home at Hill Top Farm at Near Sawrey, and other roads circle the shore and offer views down towards the lake. At Brockhole, south west of the town, is the National Park Centre.

WINDERMERE runs south and north of Bowness, 14 miles west of Kendal.

A view of Windermere from the shore by Bowness

WINDERMERE AND BOWNESS
Cumbria

The twin towns of Windermere and Bowness, the one on the hill and the other by the lake shore, have now expanded and thereby combined into a whole. Windermere, a mile from the lake, really began to develop after the railway arrived in 1847, when it became popular with rich Northern mill owners and industrialists who came here to build country houses, many of which have now become hotels. Today, Windermere is very much a tourist town, quite large, quite attractive, full of restaurants, cafés and hotels, with good walks and great views.

Bowness, down by the shore, is much older and is said to date back to Viking times. St Martin's Church dates from the late 15th century and the churchyard contains the mass grave of 47 people who drowned when a ferry sank in 1635; the first mention of a cross-lake ferry from Bowness is made as early as 1454. Attractions here, apart from the lake itself, include the Windermere Steamboat Museum, and the Model Railway Exhibition (Railrama) in Fallbarrow Road.

In spite of the inevitable influx of tourists, this centre of Bowness and Windermere is still one of the best places to start any tour of the Lake District, by steamer, car, or on foot.

WINDERMERE and Bowness lie 14 miles west of Kendal.

Windermere, the most popular lake, seen under cloud-filled skies from Summers How

Index

Page numbers in **bold** denote illustrations.

Acknowledgements

AA Picture Library: endpapers, title verso, contents, 6, 12/3, 19, 20, 21, 24, 28, 29, 33, 37(l,r), 38, 44, 45, 50/1, 52, 52, 53, 54, 56, 57, 58, 59, 60, 61, 65, 66, 67, 69(l,r), 70/1, 71,72, 73, 76/7, 78, 79, 80(l,r), 81, 83, 84, 85, 87, 89, 92, 94/5, 96, 97, 98, 99, 104, 105, 106, 108, 110, 112, 113
Janet and Colin Bord: 14,63,68
Britain on View: 32, 35, 42/3, 49, 91, 93, 95, 101, 102, 103, 107, 111
John Heseltine: title page, 9, 17, 22(t,b), 23, 25, 26
Neil Menneer: 31,39,41
Colin Molyneux: 8, 10
John Sims: 3, 4, 4/5, 7, 11
Harry Smith Collection: 27
Woodmansterne: 40